Endorsements for *Fossils*

As a child, I used to love poring over books on dinosaurs and illustrations over and over again. For a long time, I have considered it a shame that no such books exist from a biblical perspective. I am now happy to say that such a work does exist: *Fossils and the Flood* by Paul Garner. I was thrilled to see a clear presentation of a creationist understanding of earth history, and I was perhaps even more thrilled to see depictions of fantastic creatures that never make it into the public eye, from the tiny *Microdictyon* to the bulky *Moschops*. This book is a delight filled with treasures for children, families, and future scientists who want to understand paleontology from a biblical worldview.

Dr. Matthew McLain,
Associate Professor of Biology and Geology,
The Master's University, Santa Clarita, California

Fossils and the Flood is a delightful book for the young and the young-at-heart. It reminds me of the field guides and dinosaur books that captivated me as a boy and nourished my fascination with paleontology, and yet surpasses them. For in this richly illustrated and thoughtfully written book, Paul Garner leads us in a more excellent understanding of the fossil record: that it is a testimony to the great Creator and Judge of this world.

Paul Garner possesses a rare skill among scientists: he communicates technical ideas to the general audience with clarity, passion, and great skill. The result is an eminently readable book that brings the reader to a deeper understanding of the history of the world as sketched by the pages of Genesis and enriched in detail from the findings of science.

The written descriptions of the pre-Flood biomes pair with the visual depictions marvelously. There is enough detail to convey a real sense of each environment and the creatures that inhabited them. By weaving all the biomes together, the reader is transported to a world that is richly filled with living beings both familiar and exotic.

Fossils and the Flood finds itself equally useful in the field as it does in the library. The section on major fossil groups has so many helpful images of common fossils that the amateur collector should make sure to take this book to the outcrop along with their hammer, goggles, and collection bags.

Dr. Marcus Ross,
CEO, Cornerstone Educational Supply;
Fellow, Center for Creation Studies,
Liberty University, Lynchburg, Virginia

Books on fossils and geology can sometimes be hard to read but this book is certainly an exception. It is very easy to follow despite being scientifically comprehensive. And the amazing illustrations make it a joy to read. I like the way it has separate sections on historical details and answers to some of the 'big questions' such as the Flood and Noah's ark. Anyone interested in earth history must get this book.

Professor Stuart Burgess,
Professor of Engineering Design,
University of Bristol, UK

Our conception of the past is inevitably shaped by our experience of the present. *Fossils and the Flood* goes a long way in helping us to overcome that limitation, unveiling a lost world. The reconstruction Paul Garner provides is not speculation, but informed by evidence from geology understood within the secure historical framework of scripture. All this is brought to life in a book saturated with stunning and detailed artwork that is scientifically nuanced.

Deceptively easy to read, this book contains a wealth of learning that showcases the best of contemporary creationist research. It cannot fail to excite a new generation with the scientific beauty and power of a creation model—and inspire them to develop it further! If you want to know what a creation model is, and why it is worth pursuing, I can't think of a better place to begin.

Dr. Stephen Lloyd,
Pastor, Hope Church, Gravesend, Kent;
Researcher and Lecturer, Biblical Creation Trust, UK

I am committed to the theology of biblical creationism. This does, however, leave me with lots of questions in the face of the popular flood of evolutionary assumptions. This book introduces plausible explanations for the geology of a young earth. I valued the honesty and integrity of the book. It acknowledges that more research is needed in order to develop our understanding of the fossil record. This book will encourage many and intrigue others. I share the author's hope that it will inspire some to enter the world of geology and other sciences to do the hard work of research with clarity and respect, so that non-scientists like me can be built up in our faith.

Paul Spear,
Pastoral Dean,
Union School of Theology, Bridgend, UK

Paul Garner has, again, produced a much welcomed—and needed—work. This time, Garner and (illustrator) Jeanne Elizabeth provide younger readers two books in one: a comprehensive survey of early earth history and an overview of the science of paleontology. *Fossils and the Flood* provides a superbly articulated introduction to the creation model. Readers will find a detailed narrative for creation history that is accurately illustrated at every step.

Younger aspiring scientists will also find an excellent introduction to the science of paleontology—all framed in terms of creation geology, geography, and biology. The authors include an illustrated survey of every major fossil group.

Beyond this being the most accurate creation model articulation for a younger audience, I would strongly encourage older readers to not underestimate the contents. Garner skillfully weaves together nuanced threads of the creation model that are sometimes misunderstood by even mature creationist audiences. My hope is that *Fossils and the Flood* will reach a wide audience.

Dr. Neal Doran,
Professor of Biology and Director,
Center of Creation Research,
Bryan College, Dayton, Tennessee

Children and adults alike will spend hours looking at this unique and beautifully illustrated book about earth history and fossils from a biblical perspective. The book stimulates the imagination about what the earth was like before, during, and after the great Flood of Noah. The text of the book includes the most recent and up-to-date ideas of creation scientists but holds to a literal interpretation of scripture and explains it with an enjoyable writing style which the layperson will understand. Everyone needs a copy of this on their coffee table!

DR. JOHN WHITMORE,
Senior Professor of Geology,
Cedarville University, Cedarville, Ohio

Paul Garner has done the church a great service. In page after page, he unveils the astonishing beauty and diversity of God's creation from a Christian perspective. The book exudes confidence in scripture and natural science's capacity to glorify God. Accompanying Garner's prose are Jeanne Elizabeth's wonderful illustrations, which offer a feast for the eyes! I highly recommend *Fossils and the Flood* to every young believer. If you are remotely interested in science, you will devour this book—receiving delight, instruction, and encouragement with every page.

DR. HANS MADUEME,
Associate Professor of Theological Studies,
Covenant College, Lookout Mountain, Georgia

Fossils and the Flood is a magnificent and grand sweep of the wealth of discoveries that have been made in geology which give overwhelming evidence for the worldwide Flood and the rapid burial of plants and creatures across the globe. Paul Garner has made the evidence very accessible, such that the layman without specialized knowledge can understand each section. There are excellent dioramas to summarize the flora and fauna that were in existence prior to the Flood, and Garner has provided further detail of the references to these creatures at the website which accompanies this book (www.fossilsandtheflood.net). He rightly points out that there was a much greater diversity of creatures then than we have today, and that this is apparent in the fossil record.

Garner presents an excellent description of the Flood and the possible mechanics of how it came about, along with the explanation of how zones of different ecosystems of creatures may have been buried. What is especially of interest is the strong evidence of rapid burial, the end of the Flood, and the diversification of the animal kinds that came off the ark of Noah. The explanation of human diversity after the Flood and through the single subsequent ice age is very compelling. His thorough treatment of how to understand the different types of fossils will be, for many readers who are amateurs at this, a source of great inspiration as they read these final pages. This book is a heartening answer to the evolutionary philosophy with which we are so often bombarded.

Few books have given the overall picture of geology and how to interpret the rocks and fossils within a biblical context. This book is a milestone that will be referred to for many years to come.

PROFESSOR ANDY MCINTOSH,
Emeritus Professor of Thermodynamics,
University of Leeds, UK

FOSSILS AND THE FLOOD

EXPLORING LOST WORLDS WITH SCIENCE AND SCRIPTURE

by PAUL GARNER

Illustrations by JEANNE ELIZABETH

Paul Garner is a full-time researcher and lecturer for Biblical Creation Trust in the United Kingdom. He has a master's degree in geoscience from University College London, where he specialized in paleobiology. He is a fellow of the Geological Society of London and a member of the Geological Society of America, the Palaeontological Association, and the Society of Vertebrate Paleontology. His first book, *The New Creationism: Building Scientific Theories on a Biblical Foundation*, was published by Evangelical Press in 2009.

Jeanne Elizabeth studied fine art at Harrow Art College, qualifying for her National Diploma of Design. She has been employed as an illustrator in a London firm, and also accepts private commissions. She is proficient in both illustration and portraiture, using a variety of media.

FOSSILS AND THE FLOOD: EXPLORING LOST WORLDS WITH SCIENCE AND SCRIPTURE
by Paul Garner. Illustrated by Jeanne Elizabeth.

All Bible quotations are from the Authorized King James Version.

Cover illustration "Post-Flood Lakes" by Jeanne Elizabeth (see page 72); design by Joy Miller and Benjamin Kelley. Book design by Benjamin Kelley.

Title: Fossils and the flood: exploring lost worlds with science and scripture / Paul Garner (author), Jeanne Elizabeth (illustrator).

Description: Hardback edition. | Nashville: New Creation, 2021.

Identifiers: ISBN 978-0-9990409-6-6 (hardcover)

Subjects: BISAC: SCIENCE / Paleontology | RELIGION / Religion & Science | NATURE / Fossils

FOSSILS AND THE FLOOD

EXPLORING LOST WORLDS WITH SCIENCE AND SCRIPTURE

by PAUL GARNER

Illustrations by JEANNE ELIZABETH

Dedicated to my mum, Carolyn Garner.

Thank you for everything.

"For I the Lord thy God will hold thy right hand, saying unto thee, Fear not; I will help thee."

Isaiah 41:13

FOREWORD

This is a magnificently illustrated book that skillfully and systematically walks through the fossils preserved in the earth's rock record within the biblical framework of earth history centered on the Genesis Flood. Carefully crafted, easily understood descriptions introduce readers to the pre-Flood biological communities that lived in different geographic zones and were thus progressively destroyed as the floodwaters rose and buried them in the sediment layers spread across the continental fragments produced by the erupting of the fountains of the great deep splitting apart the original supercontinent. Next is a primer on fossils and how they form and are preserved, followed by a survey of all the major classified fossil groups.

Throughout, the Bible is handled robustly and the powerful scientific evidence presented that confirms the reliability, truth, and authority of the Bible. I can thus thoroughly and enthusiastically endorse this book and recommend it to children of any school age, budding young paleontologists, parents, and even grandparents. You'll learn so much about this beautifully presented fossil evidence and be superbly equipped to defend God's Word.

Dr. Andrew A. Snelling,
Director of Research,
Answers in Genesis (USA)

CONTENTS

PREFACE

This book has its origins in my childhood. As a boy I was fascinated by tomes reconstructing the history of life from the fossil record. The dioramas they contained, depicting past environments populated with long-vanished animals and plants, captivated my interest and even influenced my decision later in life to study geology.

Invariably, these books were written from an evolutionary perspective. Now as an adult and a creationist researcher I wonder, *Where are the books that can inspire similar interest in young people today, but which present the evidence of the fossil record within the biblical framework of earth history?*

There are books written for Christian families that deal with aspects of the Bible's record of early earth history, many focusing on the account of Noah's Flood. But too often they get the biblical and scientific details wrong. One common error is to inaccurately portray the size and likely shape of the ark. Another is to depict modern species on the ark (such as lions and tigers) rather than the ancestral representatives of their kinds, as would have been the case (such as the ancestral cat that, in all probability, gave rise to modern lions and tigers).

Even creationist books that try to rectify these mistakes can still be at odds with the findings of cutting-edge creationist research in other respects. For instance, they may portray Noah and his family as the typical Bible story characters—too modern in physical appearance and dress. Or they may include dinosaurs on the ark but overlook all the other kinds of extinct animals that must have been represented.

So this book is my attempt to present an informed and coherent synthesis of the best biblical and scientific scholarship within creationism on the theme of the fossil record. In my efforts to get the details right, I consulted with several experts. In particular I am grateful to Dr. Kurt Wise (PhD invertebrate paleontology), Dr. Andrew Snelling (PhD geology), Dr. Todd Wood (PhD biochemistry), and the late and much-missed Dr. Roger Sanders (PhD botany) for their insights, comments, and criticisms while I was preparing this manuscript. I know that the book has benefited enormously from the generous time and effort they expended in helping me, though this should in no way be taken to suggest that they agreed with everything I wrote. Any remaining errors, of course, are my sole responsibility. I am also grateful for the editorial input of Thomas Purifoy, Mike Matthews, and Ben Kelley, whose suggestions greatly improved the finished manuscript. Ben also deserves recognition for his stunning work on the book design.

I am also immensely fortunate to have had the opportunity to work with such a talented illustrator as Jeanne Elizabeth. Though she was unused to taking direction from "difficult" scientists, she threw herself into this project with great enthusiasm and aplomb! I am sure that readers will agree that Jeanne's marvelous paintings are the making of this book and give it tremendous visual appeal. This is her book every bit as much as it is mine, perhaps more so.

Since I intended for this book to be read by lay people and not scholars, I have resisted the temptation to use footnotes in the text. Readers who would like to delve into the evidence (both biblical and scientific) on which the book's claims are based have a resource in the notes and references listed at the back of this book and, even more extensively, on the web page that accompanies this book (see "Recommended resources" and www.fossilsandtheflood.net).

I hope you enjoy what you are about to read–and all the wonderful illustrations. Perhaps there will even be a young reader whose imagination about the past is captured by what they read and see, just as mine was all those years ago. How gratifying it would be to know that a researcher of tomorrow decided to take up the scientific quest because they picked up this book.

Paul Garner

"The works of the LORD are great, sought out of all them that have pleasure therein."

Psalm 111:2

WHAT THIS BOOK IS ABOUT

This book is about the worldwide Flood that took place in the days of Noah.

We know quite a lot about the Flood because it was recorded for us in the book of Genesis, the first book of the Bible. We can trust this written record because, as 2 Timothy 3:16 tells us, "All scripture is given by inspiration of God," and we know that God is always truthful. Numbers 23:19 says, "God is not a man, that he should lie" (see also 1 Samuel 15:29; Romans 3:4; Titus 1:2).

But the Bible does not answer all of our questions. The Flood also left evidence that can be collected and studied by scientists. This scientific evidence gives us additional information about how the Flood happened. We can even begin to work out what the world was like before the Flood. Much of this evidence comes from the study of fossils.

With the help of God's Word and clues from our scientific studies we can piece together a fuller picture of this remarkable time in the earth's history. This book tries to do just that.

The Bible is our most important source of information about the Flood because it was inspired by God and is therefore completely reliable.

God has given us the tools of scientific investigation so that we can discover things not directly revealed in the Bible, but this information is never as reliable as the Bible itself.

Of course, scientific answers are never final. Scientific ideas change as new evidence comes to light and some may turn out to be completely wrong. But the Bible's historical record of the Flood is true and will never change.

This book is our attempt to take the puzzle pieces before us and put them together into a meaningful picture. But you should remember that some of the details are simply our best guess right now.

HOW THIS BOOK IS ORGANIZED

This book is divided into six main sections.

The first section considers what the Bible tells us about *the early history of the earth* as well as the kinds of insights that we can glean from scientific investigation.

The second section looks in detail at the old world—*the world before Noah's Flood*. It seeks to reconstruct from the biblical and scientific clues the major communities of plants and animals that existed on the earth at that time.

That world was destroyed by the Flood, however, and the third section of the book describes *how the Flood unfolded*, transporting and burying those created biomes to produce the fossil record.

A new world emerged after the Flood, and the fourth section explores how the animals repopulated the devastated earth, and *how the earth itself recovered* from the catastrophe.

The fifth and sixth sections of the book look in greater depth at what we know about *fossils and the fossil record*, and how we know it. These sections explain how fossils formed, and describe the major groups of fossil organisms.

You may find it helpful to consult the last two sections when reading the earlier parts of the book. For example, when we describe the salty, hot-water reefs of the old world, you may want to turn to the pages on microfossils to find out more about stromatolites.

The book concludes with some *recommendations for further study*, including books, DVDs, periodicals, and websites, and a *glossary* of significant terms, which appear in bold at key places in the text.

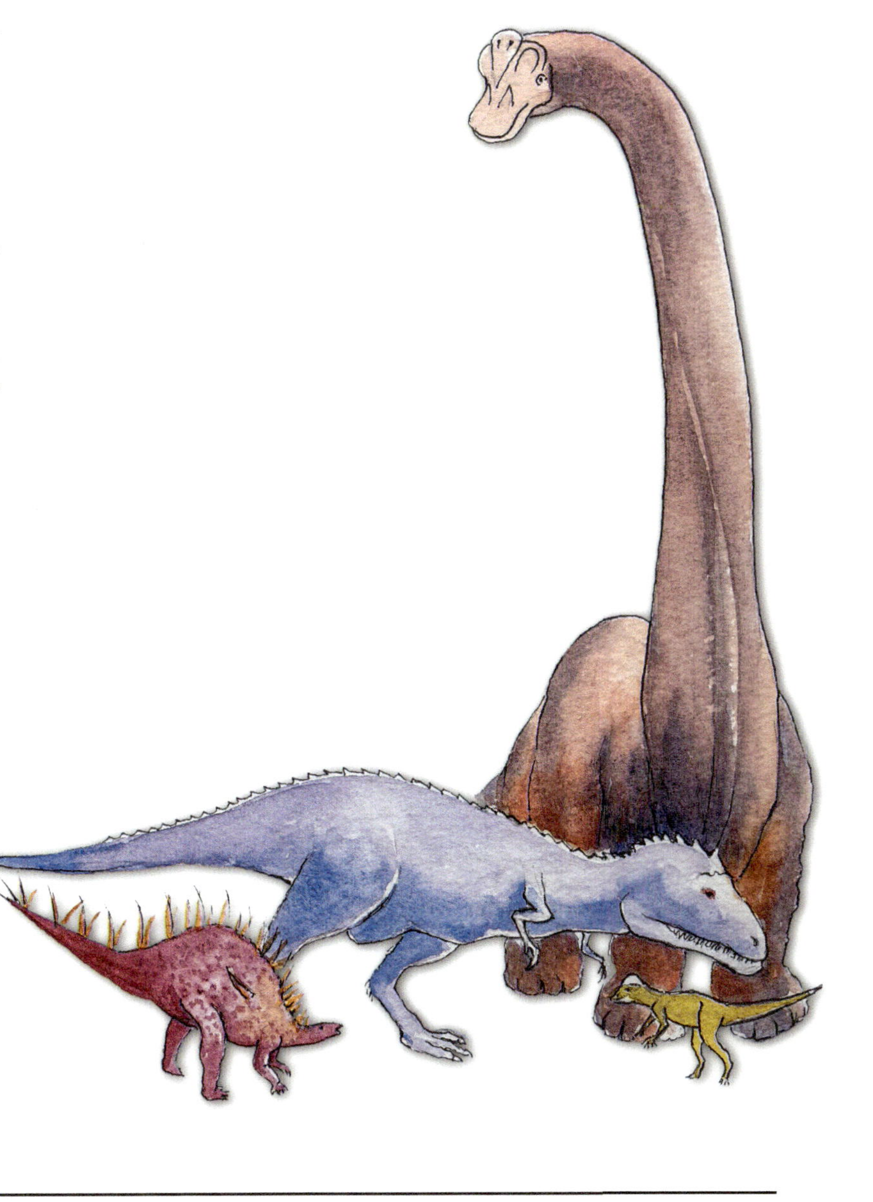

1

THE EARLY HISTORY OF THE EARTH

The Bible gives us a clear, historical framework for the early history of the earth, while science is a God-given tool that allows us to work out many of the details of these historical events. By studying both the Bible and science, we can gain a more complete picture of the earth's early history, including Creation and the Flood.

WHAT DOES THE BIBLE TELL US?

The Bible gives us a clear outline of the early history of the earth.

It tells us that God made all things in six days only thousands of years ago (**Creation**). But this was soon followed by the fall into sin of the first man, Adam, which brought death and corruption into the world (**Fall**).

By the time of Noah, the world had become so corrupt and full of violence that God judged it with a worldwide deluge (**Flood**). During the Flood, humans, birds, and land animals were preserved on the ark.

After the Flood, human languages were confused at Babel, causing the people to scatter across the face of the earth (**Babel**). Also, human lifespans rapidly declined.

We will consider each of these events in more detail later. This God-given outline of the earth's history is important because it guides the way we think about the scientific evidence throughout this book.

WHAT CAN WE LEARN FROM SCIENTIFIC RESEARCH?

The Bible gives us a true and reliable outline of the earth's early history. But God gives us the exciting task of using scientific research to find out many of the details of these events.

We can discover some things about the history of our world by studying living creatures (the science of **biology**).

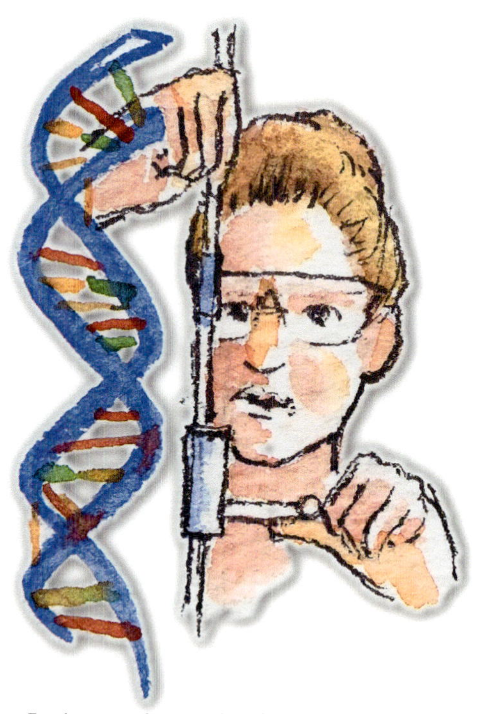

Biology is the study of living organisms.

Other clues come from investigating rocks, minerals, and the structure of the earth (the science of **geology**).

Geology is the study of the earth.

Still more can be learned by studying creatures that lived in the past, now preserved as **fossils** in the earth's rocks (the science of **paleontology**).

Paleontology is the study of fossilized organisms.

Bible-believing scientists carry out field research, conduct laboratory experiments, and develop scientific theories in each of these areas, guided by the framework of Creation, Fall, Flood, and Babel given in scripture.

SOME BIBLE BELIEVING SCIENTISTS OF TODAY

Biologist Dr. Gordon Wilson

- PhD in environmental science
- Senior fellow of natural history, New Saint Andrews College, Idaho

Dr. Wilson says: "We are studying the direct handiwork of God, and it gives us insight into His creative and artistic character, so biology is part of theology."

Geologist Dr. Steven Austin

- PhD in sedimentary geology
- President of the Creation Geology Society

Dr. Austin says: "The fact is that geologic features form rapidly and not over millions of years. The geologic evidence is entirely consistent with the biblical timescale."

Paleontologist Dr. Matthew McLain

- PhD in earth sciences
- Associate professor of biology and geology, The Master's University, California

Dr. McLain says: "From the tiniest, ornately-crafted diatom to 80-ton, 100-foot-long sauropod dinosaurs, the fossil record puts God's glory on display in a magnificent way."

PUTTING THE PUZZLE PIECES TOGETHER

The goal of Bible-believing scientists is to reconstruct the early history of the earth in a manner that is consistent with both the historical framework in the Bible and the scientific evidence.

That task is like assembling a complex jigsaw puzzle. Scripture provides the edge pieces that set the boundaries, as well as some crucial inside pieces. Science provides many additional pieces. But, unlike a jigsaw puzzle, we will never have all the pieces and there is no box to show us what the completed picture should look like. Despite these challenges, a coherent picture is emerging from this research effort.

The most significant geological event recorded in the Bible is the Flood of Noah—an event that must have left evidence in the rock record, as we will explain later in the book. This suggests that we can divide the earth's rock layers, and the fossils they contain, into those that were deposited before the Flood, those that were deposited during the Flood, and those that were deposited after the Flood.

In this book we refer to the world before the Flood as *the old world*, and the world after the Flood as *the new world*. In the next section, we will explore the old world, and seek to reconstruct it from the biblical and scientific clues available to us—with particular reference to its origins, its age, its physical features, and its animals and plants.

2

THE OLD WORLD

Clues from the Bible and science allow us to reconstruct the world before Noah's Flood—its continents and oceans and its communities of plants and animals. Many of these communities are now extinct, but we can find out about them by studying their remains in the fossil record.

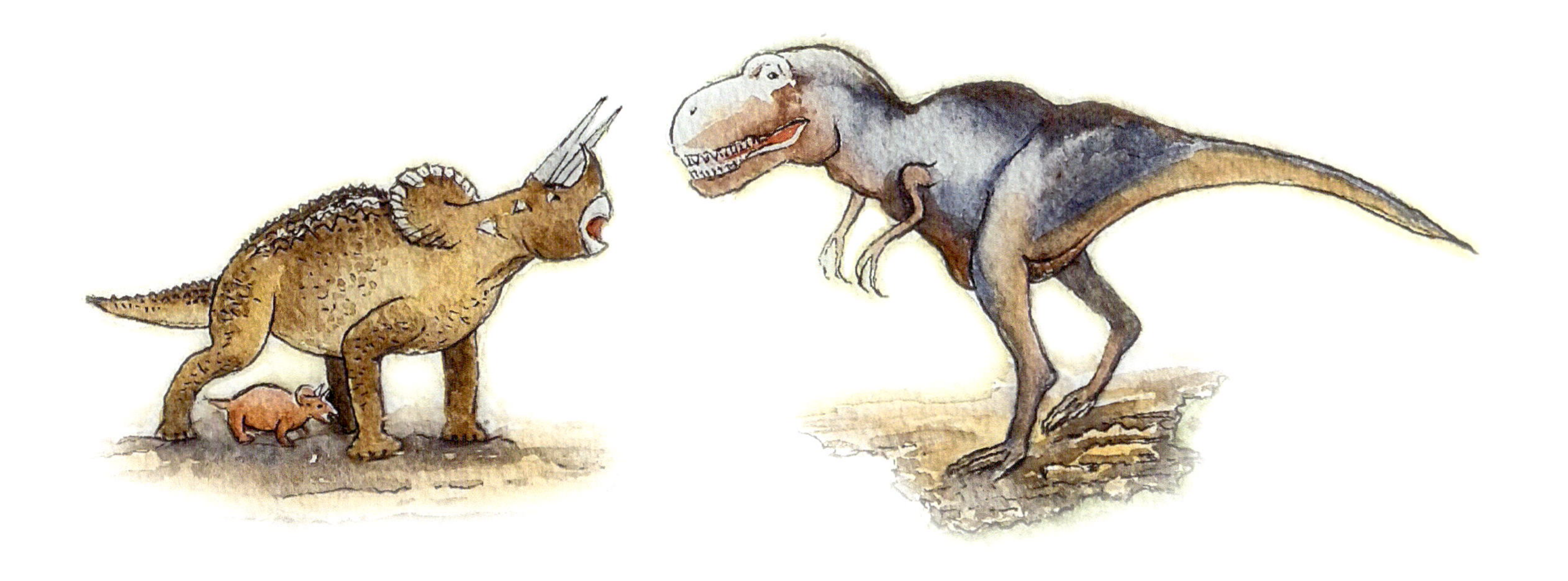

CREATION WEEK

We begin our **reconstruction** of the earth's history by considering what the Bible says about the earth's origins. It is important to start with the Bible because God the Creator inspired its human authors to write down exactly what he wanted them to, without any mistakes or errors.

The Bible's account of Creation (Genesis 1:1–2:4) tells us that God made the heavens, the earth, and everything in them in just six days:

On the first day God created the heavens and the earth covered in water, and then separated the light from the darkness.

On the second day he separated the waters above the heavenly expanse from the waters below the heavenly expanse.

▼ On the third day he separated the dry land from the seas and made the plants.

▲ On the fourth day he created the sun, moon, and stars.

▲ On the fifth day he created the flying creatures and the swimming creatures.

▲ On the sixth day he created the land animals and the first people—Adam (from the dust) and Eve (from Adam's side).

▼ By the end of the sixth day, Creation was finished. And so on the seventh day God rested from all that he had done.

HOW LONG AGO DID CREATION TAKE PLACE?

The Bible also gives us information that allows us to work out how long ago Creation took place.

Two chapters in Genesis provide important clues: Genesis 5, which traces the family line from Adam to Noah, and Genesis 11, which traces the family line from Noah to Abraham. By adding up the ages in these family lines we can in principle work out when Adam was created.

WERE THE CREATION DAYS ORDINARY DAYS?

There are many reasons to think that the days of Creation were ordinary days, each about twenty-four hours long:

1. Genesis 1:5 uses the Hebrew word *day* to describe the daylight portion of a day and the entire light/dark cycle—in other words a normal day.
2. Each day of Creation is given a number. Elsewhere in the Old Testament this always means a normal day.
3. Each day consists of an evening and a morning. Again, this always refers to a normal day elsewhere in the Old Testament.
4. Genesis 1 does not use other Hebrew words for time (e.g. *olam*—meaning *antiquity* or *eon*), which could have more clearly conveyed the idea of long or indefinite creation-days.
5. Exodus 20:8-11 draws a parallel between the Creation week and our week. We are to remember the Sabbath day and keep it holy because God created in six days and rested on the seventh.

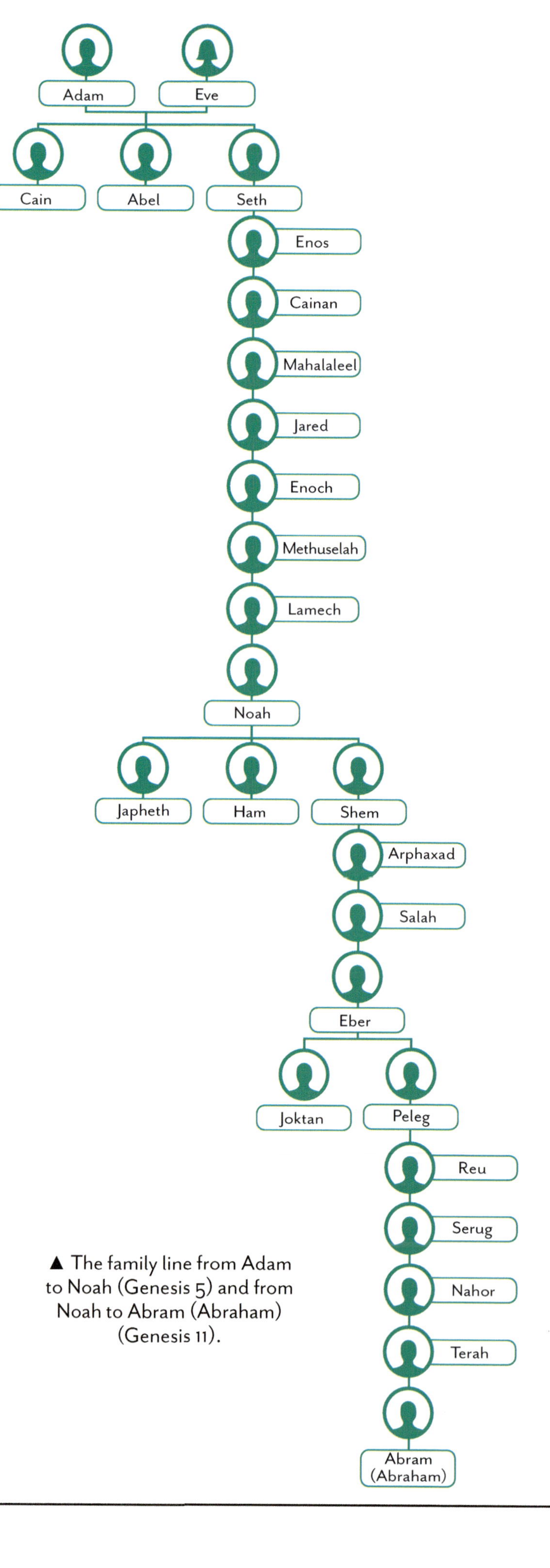

▲ The family line from Adam to Noah (Genesis 5) and from Noah to Abram (Abraham) (Genesis 11).

The task is complicated somewhat by the fact that different manuscripts of the Old Testament give different numbers. On the one hand, the standard Hebrew text of the Old Testament—the Masoretic text—suggests that Adam was created about 2,000 years before Abraham. On the other hand, the Greek text of the Old Testament—the Septuagint—suggests this period may have been about 3,400 years. Most scholars favor the Masoretic text, though some prefer the Septuagint. The difference between them is small, however, when both are contrasted with the long ages of conventional **chronology**.

Since we know that Jesus lived about 2,000 years ago and Abraham about 2,000 years before that, we can sum the ages to estimate that Creation must have taken place about 6,000 years ago (according to the Masoretic text) or 7,400 years ago (according to the Septuagint text).

Many people (including most scientists) do not accept such a recent date for Creation. They think that the world is billions of years old, and they appeal to dating methods based on the decay of radioactive atoms to support a much longer time scale. We can see from below, however, that there are many problems with these methods.

RADIOMETRIC DATING AND ITS ASSUMPTIONS

Radiometric dating uses naturally occurring radioactive elements to date rocks and minerals. Unstable, radioactive parent atoms decay through a series of intermediate steps until stable, non-radioactive daughter atoms form. The accumulation of daughter atoms allows the sample's age to be calculated.

Three assumptions must be made:

1. The rock or mineral contained a known quantity of daughter atoms in the beginning.
2. The amounts of parent and daughter atoms have not been altered by anything besides radioactive decay.
3. The rate of decay has been constant.

But biblical creationists think there are good reasons to challenge each of these assumptions:

1. The original quantity of daughter atoms is often uncertain.
2. Many processes besides radioactive decay can alter the amounts of parent and daughter atoms in a rock or mineral.
3. There is evidence that radioactive decay rates were higher during Creation and the Flood than today, which would make radiometric dates too old.

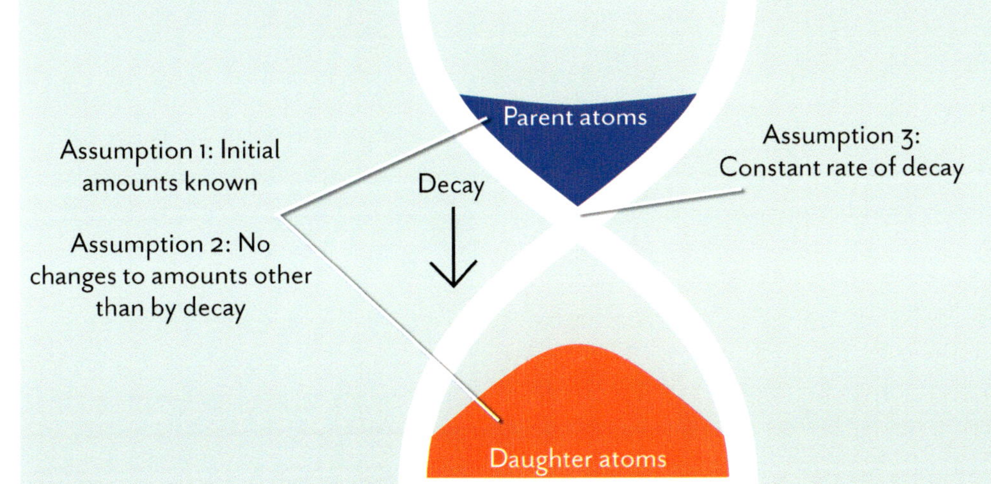

◀Geologists estimate the ages of rocks and minerals using the accumulation of stable, non-radioactive daughter atoms from the decay of unstable, radioactive parent atoms. Each parent has its own characteristic half-life—defined as the time it takes for half of a given amount of parent to decay into daughter. However, three questionable assumptions must be made.

Furthermore, there is much scientific evidence that points to a young world. This evidence includes:

Short comet lifetimes: **Comets** are mostly ice and dust, and they lose material every time they orbit close to the sun. They cannot survive more than about 10,000 years. It is hard to explain how the number of comets in our solar system could have been sustained for billions of years.

Saltiness of the oceans: Every year, rivers and other sources wash large amounts of salt into the oceans, and most of it builds up there. If the oceans were billions of years old they ought to be much saltier than they are today.

Sediment buildup in the oceans: Rivers also carry sediments into the ocean. The average thickness of the sediments on the seafloor is less than 1,300 feet (400 meters). The oceans ought to have been filled many times over in hundreds of millions of years.

Wearing down of the continents: Water and wind **erosion** is wearing away the surface of the land. If the continents were billions of years old they ought to have been worn down to sea level hundreds of times—but we still have high-standing continents that appear not to have completed even one such cycle.

THE CREATED WORLD I. CONTINENTS AND OCEANS

What was the world like when God made it? This is not an easy question to answer but the Bible and geology provide some important clues.

Genesis 1:9-10 tells us that on the third day of Creation, God gathered the waters into one place and made the dry land appear. He called the waters "seas" and the dry land "earth". So it is reasonable to conclude that the world back then had oceans and continents much as today.

But in many other respects the world before the Flood was very different from today's world. We know this because the Bible tells us that the Flood destroyed the old world in the days of Noah (2 Peter 3:6). The world we see today bears the marks of that awesome judgement.

Clues preserved in the earth's rock layers indicate that most of the earth's landmasses were originally gathered together to form one super-sized continent. Scientists have named this supercontinent **Rodinia** (meaning *motherland*). This **supercontinent** was probably made up of large areas of land separated by shallow seas and waterways.

During the Flood this supercontinent broke apart and the pieces moved around. By studying the continental fragments that remain—fragments that geologists call **cratons**—scientists can piece together how those fragments might originally have been arranged.

Putting the broken and scattered pieces of the puzzle back together is difficult. But it seems likely the continent we now know as North America was near the center of the original supercontinent, with Australia and East Antarctica along its western edge.

RECONSTRUCTING RODINIA

Geological evidence indicates that the earth's continents are not fixed but moved around in the past. Scientists study clues in the rocks to work out how the continents used to be arranged. Rodinia was reconstructed by matching up rock layers around the world as well as by studying magnetic minerals in the rocks that record the latitude at which the rocks originally formed. There is, however, disagreement about the details.

◀ One possible reconstruction of Rodinia, the pre-Flood supercontinent. It is important to note that some continental pieces have very different positions in different reconstructions. This reconstruction will almost certainly have to be revised with additional study.

THE CREATED WORLD II. LIVING THINGS

Another difference between today's world and the old world was in the diversity of living things that the old world supported.

At the time of Creation, God filled the world with an astounding array of living things. He made **bacteria**, **algae**, **fungi**, **plants**, and **animals**—and within each of these groups he created even more diversity.

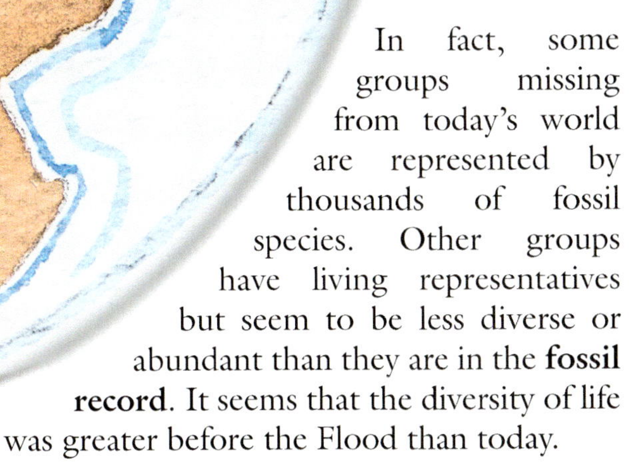

In fact, some groups missing from today's world are represented by thousands of fossil species. Other groups have living representatives but seem to be less diverse or abundant than they are in the **fossil record**. It seems that the diversity of life was greater before the Flood than today.

At the time of Creation, God made different environments to provide homes for all these creatures, just as in today's world there are grasslands, forests, deserts, lakes, and seas. These major habitats—each with its own climate and populated by a distinctive community of plants and animals—is called a **biome**. Each biome provided many ecological niches, and God made creatures in a bewildering range of sizes, shapes, and diets to fill these niches.

Deserts, grasslands, and tropical rainforests are examples of modern biomes. Each biome has its own climate and is populated by a distinctive set of plants and animals.

God's extraordinary design went even further. Today, the same biome can be found in many different places on the earth's surface, yet it is populated by a different set of plants and animals in different locations. This allows an even greater variety of creatures to live on the earth.

For example, grassland biomes include the North American prairies, the South American pampas, the African savanna, the Asian steppe, and the Australian plains—yet each supports a distinctive set of animals. In the beginning, it appears that God did something similar by creating the same environment in more than one place on the earth's surface in order to maximize the amount of diversity that could be supported.

Some of the biomes originally created by God are now extinct, along with the creatures that lived in them. We can find out about these biomes, however, by studying the remains left behind in the fossil record. Over the next few pages we will take a close look at some of the biomes in the old world before the Flood. We will begin with an extraordinary biome that was floating over the deep ocean, before moving on to shallow ocean biomes, coastal biomes, and land biomes.

The presence of the same biome in many places on the earth's surface allows for an even greater range of plants and animals to be supported. For example, grasslands are found on several of today's continents and each provides a home for particular herbivores:

North American prairie with an American bison.

South American pampas with a llama.

African savanna with a springbok.

Asian steppe with a saiga antelope.

Australian plain with a kangaroo.

WHAT WAS EARTH'S CLIMATE LIKE BEFORE THE FLOOD?

Fossilized plants—and especially their growth rings—provide some important insights into the climatic conditions of the pre-Flood world.

The range of plant types buried in Flood sediments suggests that they must have grown in a variety of climates before the Flood. Temperatures overall were probably higher than at present, and the temperate zones probably extended much farther towards the poles.

Some fossil trees buried during the Flood display growth rings with evidence of seasonality, late frosts, and severe droughts, indicating that the higher latitudes experienced distinct seasonal changes.

▼ Growth rings in fossilized trees from the Purbeck Group (Upper Jurassic to Lower Cretaceous) of Dorset, England, indicate they must have grown in a strongly seasonal environment.

◀ The famous "Fossil Forest" of Lulworth, Dorset, England. This photograph shows the hollows left where fossilized tree stumps once stood.

THE FLOATING FOREST

The first of the major, pre-Flood biomes we will consider is the **floating forest**, a vast, thick mat of vegetation that floated over the deep, open ocean. Judging by the amount of fossilized plant material it left behind, this mat was probably the size of an entire continent.

At the edges of the floating forest grew small plants without true roots or leaves. They were well suited for life close to the water.

Farther into the floating forest there were small to medium-sized branching and bush-like plants. Some resembled ferns, while others were more like clubmosses. They were a bit less dependent on water, with tougher stems and extra vessels to carry water.

Large trees, including clubmosses and other lycopsids, grew in the center of the floating forest. Here the mat was thick enough to support trunks up to 100 feet (30 meters) tall and 6 feet (2 meters) around.

Many strange animals inhabited this floating forest. Scurrying through the leaf litter were insects, spiders, scorpions, and cockroaches. Giant millipede-like animals made wide trackways resembling tire marks.

The air above the floating forest buzzed with winged insects, including *Meganeura*, a giant dragonfly-like creature with a wingspan up to 30 inches (76 centimeters) from tip to tip.

Other strange creatures lurked in shallow pools on the floating mat. They had feet and legs (not fins) but in other ways were quite fish-like. These animals seem to have been well suited for life

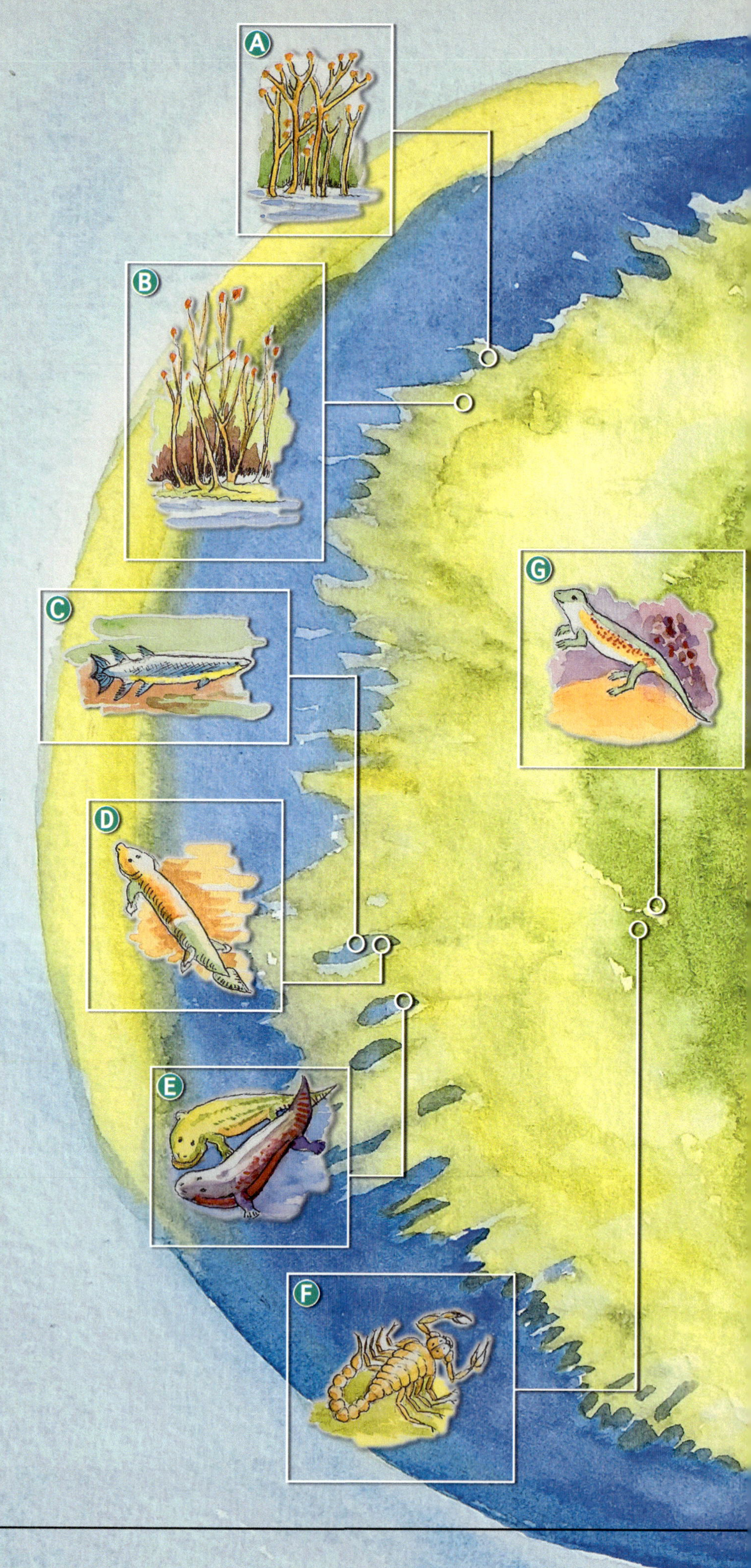

in an ecosystem that was neither fully aquatic nor fully terrestrial.

Towards the center of the floating forest were animals more suited to life away from the water, including reptiles such as *Hylonomus*.

A *Cooksonia.*
B *Rhynia* and *Zosterophyllum.*
C *Eusthenopteron.*
D *Tiktaalik.*
E *Acanthostega* and *Ichthyostega* (not to scale).
F *Pulmonoscorpius.*
G *Westlothiana.*
H *Psaronius, Neuropteris,* and *Calamites.*
I *Lepidodendron* and *Sigillaria.*
J *Meganeura* and *Arthropleura.*
K *Hylonomus.*
L *Cordaites.*

THE GIANT CLUBMOSSES OF FOSSIL GROVE, GLASGOW, SCOTLAND

In a building in Glasgow's Victoria Park is the famous Fossil Grove, a group of eleven large tree stumps belonging to giant clubmosses. They have been preserved exactly where they were found back in 1887. In fact, the stumps are not the remains of the actual trees, but rather the result of sediment filling the hollow insides of the trees. The outermost layer of woody material rotted away to leave these extraordinary natural casts—a mute testimony to the long-lost pre-Flood floating forest.

PLANTS THAT GREW IN WATER

Evidence that some of these trees grew in water, not in soil, can be seen by examining their anatomy. For instance, lycopsids had roots called ***Stigmaria*** with secondary rootlets arranged around the roots like spokes around a wheel. This type of root pattern is found only in water plants. Also, the trees, roots, and rootlets appear to have been hollow and filled with air when the plants were living—an ideal design for floating in water.

▲ Fossil specimen and ▼ life reconstruction showing the spoke-like-arrangement of *Stigmaria* rootlets.

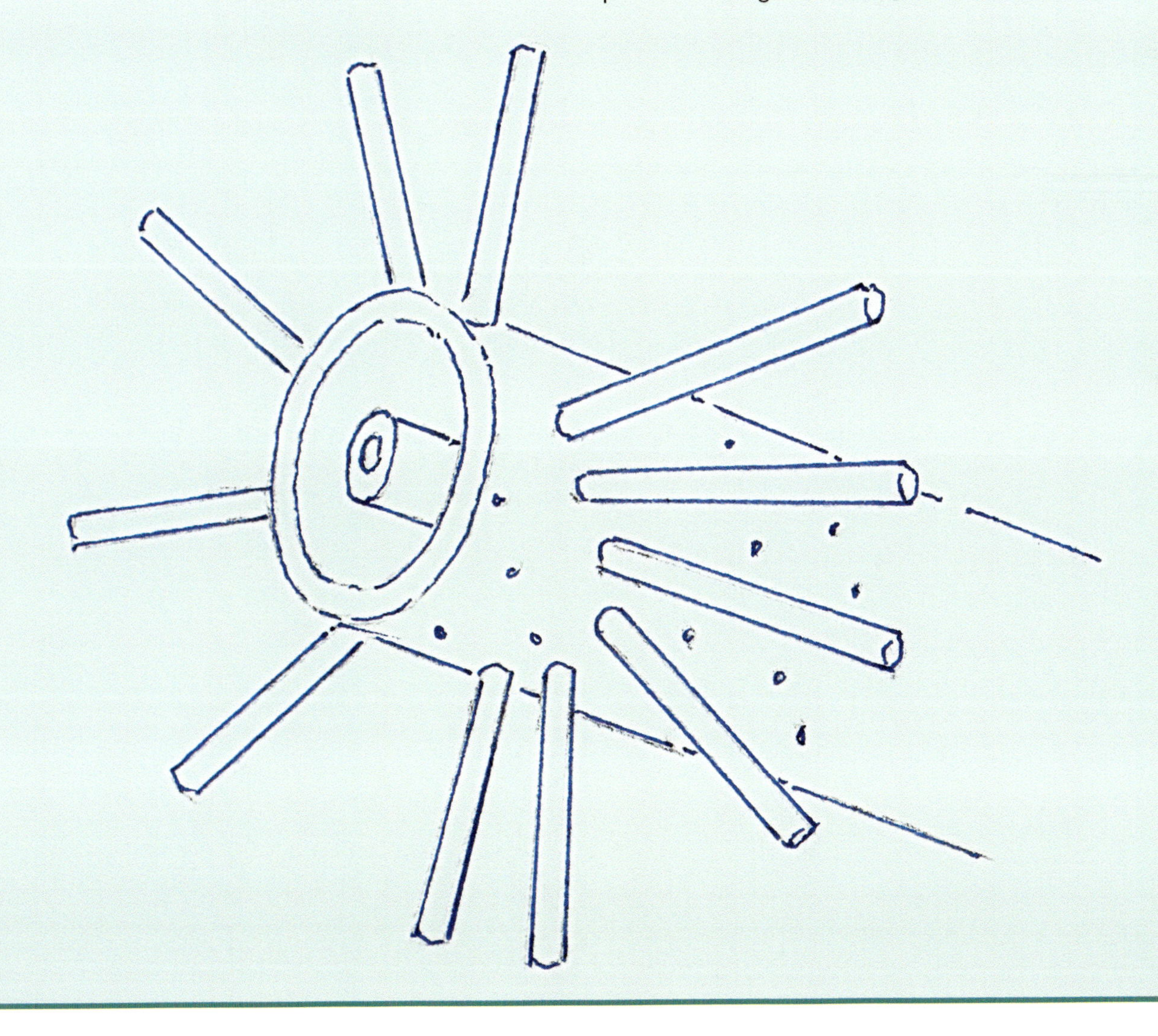

SALTY, HOT-WATER REEFS

Another unusual pre-Flood biome was a continent-fringing stromatolite **reef**. The word *stromatolite* comes from the Greek words *stromata* meaning *layers* and *lithos* meaning *rock*. Stromatolites are layered mounds or columns built by microbes trapping sediment particles (such as sand grains).

It seems that parts of the pre-Flood supercontinent were surrounded by a shallow-water marine shelf hundreds of miles wide. The part of the shelf farthest from the land was raised up by hot, buoyant rocks below so that it was close to sea level. The raised shelf margin acted as a kind of barrier between the open ocean on one side and a wide, shallow marine lagoon on the other.

The hot rocks below the shelf edge fed a system of hot springs bringing warm, salty waters to the surface, conditions that were just right for stromatolites to grow. Consequently, this outer part of the marine shelf supported an enormous stromatolite reef, unlike any biome that exists today.

The shallow water allowed plenty of light to reach the microbes building the stromatolites and the salty conditions inhibited animals that might have grazed on them. This meant that the stromatolites could grow abundantly and to large sizes.

There are places where stromatolites can be found growing today, such as in the warm, salty waters of Shark Bay in Western Australia. But these are a poor reflection of a biome that was much more extensive and supported a much greater diversity of stromatolite-building microbes before the Flood.

STROMATOLITES IN THE GRAND CANYON, ARIZONA

One place where you can see fossilized stromatolites that were growing before the Flood is in the Precambrian Kwagunt Formation in the Grand Canyon, Arizona, USA. An extensive layer within this formation contains hundreds of mushroom-shaped stromatolites, each about 8.2 feet (2.5 meters) tall. They probably grew in the shallow, warm, and salty waters of a stromatolite reef that fringed the margins of the pre-Flood supercontinent.

▼ The stromatolite bed in the Precambrian Kwagunt Formation in the Grand Canyon, Arizona.

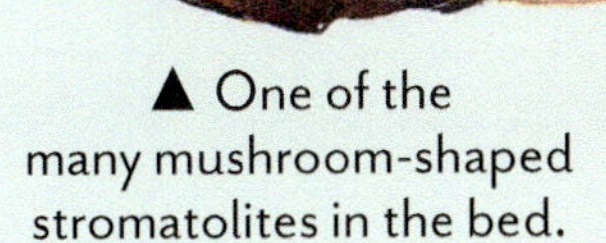

▲ One of the many mushroom-shaped stromatolites in the bed.

▲ This stromatolite cross section on display at the Grand Canyon demonstrates its layered structure.

LIVING ON THE MARINE SHELF I. THE EDIACARANS

The shallow waters of the marine lagoon behind the stromatolite reef were home to many creatures.

In the deepest waters closest to the reef there were large, soft-bodied organisms called **Ediacarans**, named after the Ediacara Hills in South Australia where their fossils were discovered. Their fossils are mostly preserved in sandstone, so they probably lived in a sandy environment.

Most of the Ediacaran creatures had bodies with a quilted appearance, and some seem to have lived in dense colonies. A few might be classified in modern groups if they were living today, but most were unlike any creatures in the modern world.

Spriggina was a worm-like organism with a head shaped like a horseshoe. *Cyclomedusa*, an enigmatic creature, had a circular body and concentric rings. *Charniodiscus* had an elongated frond-like body attached to the seafloor, while *Bradgatia* had a cabbage-like appearance with six or more fronds radiating from a central anchor point. *Tribrachidium* was a disc-shaped creature with a body divided into three parts, and *Dickinsonia* had a broad, flat body with segments arranged in a radial pattern.

Similar fossils are now known from many places around the world, including Great Britain, Namibia, and Newfoundland.

A *Charniodiscus.* C *Cyclomedusa.* E *Bradgatia.*
B *Dickinsonia.* D *Spriggina.* F *Tribrachidium.*

EDIACARANS AT MISTAKEN POINT, NEWFOUNDLAND

▼ Multiple rock layers bearing Ediacaran fossils are exposed along the shoreline at Mistaken Point in Newfoundland.

▲ A spindle-shaped fossil called *Fractofusus* on one of the bedding planes.

Mistaken Point, on the southernmost tip of the Avalon Peninsula in Newfoundland, is famous for its Ediacaran fossils. Frond-like, bush-like, and spindle-like forms are found in large numbers on the exposed rock surfaces. There are also abundant disc-shaped forms. Some are similar to fossils from the Charnwood Forest in England but others have not been found anywhere else in the world. These soft-bodied animals probably lived in moderately deep water and were preserved at the beginning of the Flood when they were smothered by volcanic ash carried by submarine currents.

▼ A digital representation showing what *Fractofusus* may have looked like.

LIVING ON THE MARINE SHELF II. SMALL SHELLY CREATURES

Farther into the marine lagoon there were lime-rich muds in which many small shelly creatures lived.

Some had coiled shells while others were cone- or tube-shaped. Few of the shells are more than 0.5 inches (1 centimeter) long. We cannot be certain what kinds of animals lived in these shells, but they may have been worms or mollusks. Some may even represent parts of the skeletons of bigger animals.

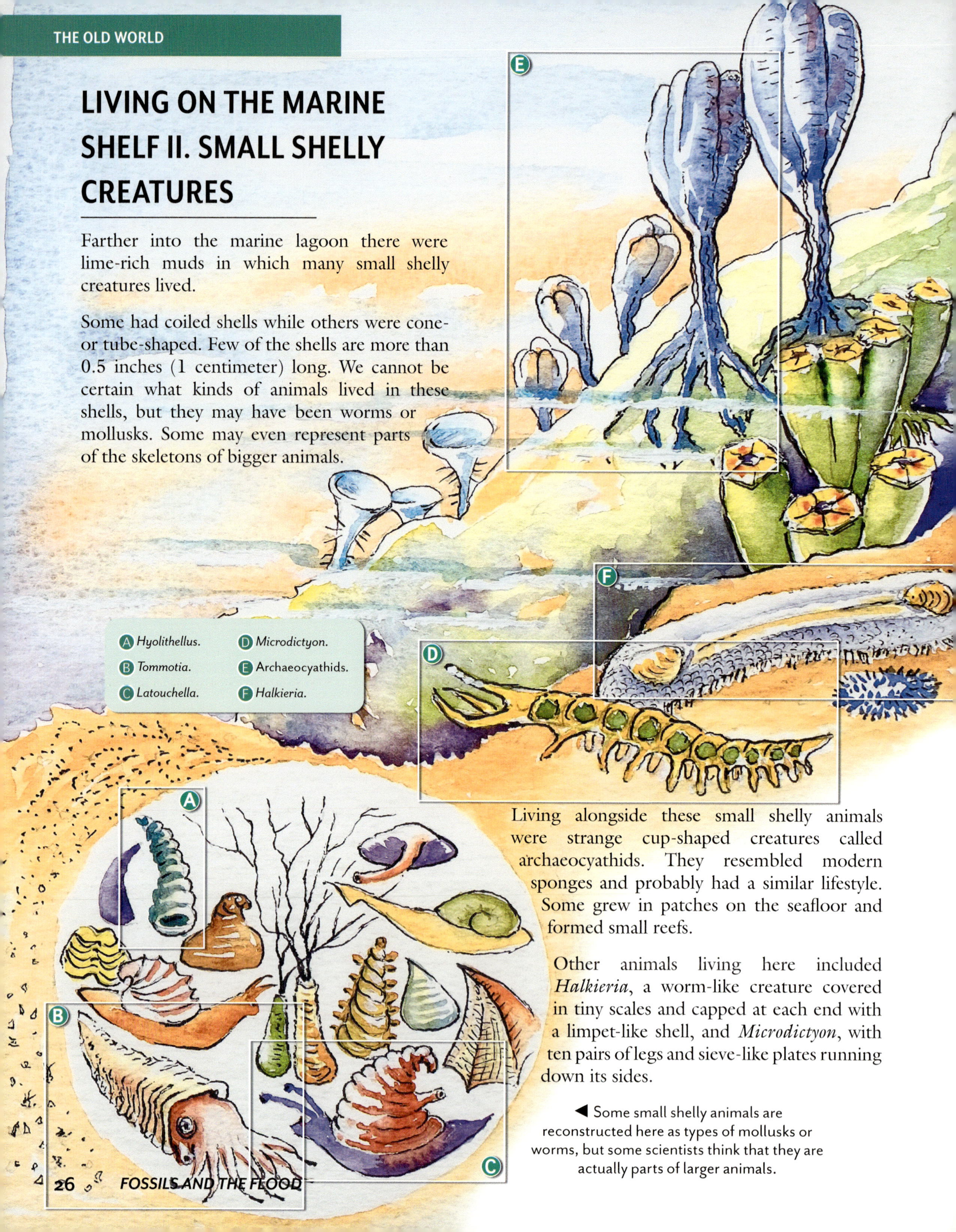

Living alongside these small shelly animals were strange cup-shaped creatures called archaeocyathids. They resembled modern sponges and probably had a similar lifestyle. Some grew in patches on the seafloor and formed small reefs.

Other animals living here included *Halkieria*, a worm-like creature covered in tiny scales and capped at each end with a limpet-like shell, and *Microdictyon*, with ten pairs of legs and sieve-like plates running down its sides.

◀ Some small shelly animals are reconstructed here as types of mollusks or worms, but some scientists think that they are actually parts of larger animals.

SMALL SHELLY CREATURES OF THE SIBERIAN PLATFORM

The fossilized remains of these small shelly animals were first discovered in rocks exposed along the Lena and Aldan Rivers in a remote and inaccessible part of Siberia.

Since then, similar fossils have been found in many other places around the world, including China, India, Canada, England, France, and Australia. Scientists sometimes refer to these fossils as the Tommotian fauna, after the subdivision of Cambrian rocks in which they were first identified.

▼ Spectacular pillars of rock along the Lena River in the Sakha Republic (Yakutia) are composed of Lower Cambrian (Tommotian) sediments yielding fossils of the small shelly creatures.

LIVING ON THE MARINE SHELF III. THE ATDABANIAN ANIMALS

The part of the marine lagoon nearest the land teemed with yet more strange animals.

The giant predatory arthropod *Anomalocaris* and the fish-like vertebrate *Myllokunmingia* swam above the seafloor, while trilobites such as *Olenellus* lived on the muddy sea bottom or burrowed into it.

On the seabed, worms such as *Paraselkirkia* waited for passing prey in their U-shaped burrows while clusters of brachiopods such as *Longtancunella* strained tiny food particles from the seawater. *Quadrolaminiella* had a long vase-shaped body and was probably a kind of sponge.

One oddity was *Hallucigenia*, with its multiple legs and shoulder spines. It may have been a type of spiny velvet worm. *Paucipodia* was similar but lacked the paired spines on its back.

There were also echinoderms, but not like the starfish and sea urchins that we are familiar with. The enigmatic *Helicoplacus* was shaped like the bob on the end of a plumb line and covered with small armor plates. It did not have the five-fold symmetry of modern echinoderms.

Another strange creature was *Eldonia*, with its soft, disc-shaped body and tentacles. Although it somewhat resembled a jellyfish, it is not known exactly what kind of animal *Eldonia* was.

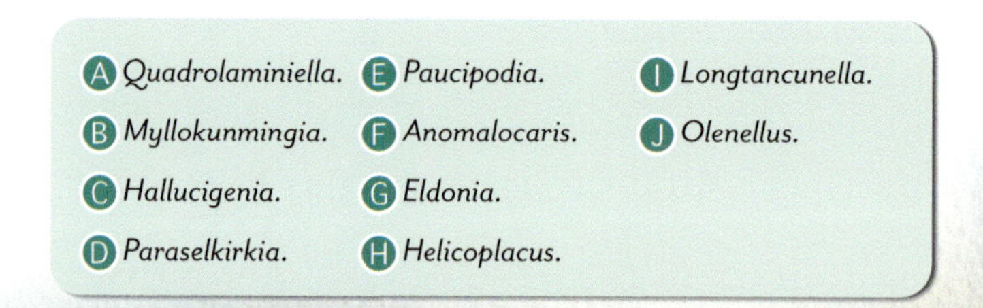
A *Quadrolaminiella.*
B *Myllokunmingia.*
C *Hallucigenia.*
D *Paraselkirkia.*
E *Paucipodia.*
F *Anomalocaris.*
G *Eldonia.*
H *Helicoplacus.*
I *Longtancunella.*
J *Olenellus.*

ATDABANIAN FAUNA OF CHENGJIANG, SOUTHERN CHINA

The Chengjiang fossil deposit in southern China preserves a remarkable array of animals that populated this part of the shallow marine shelf before the Flood. More than 150 species have been found at this site, about 60 of them arthropods. Others include worms, sponges, brachiopods, and what seem to be vertebrates. Even the soft-bodied animals are well preserved, suggesting that they were rapidly buried by catastrophic sediment flows during the Flood.

▶ The Lower Cambrian (**Atdabanian**) rocks of Chengjiang in Yunnan Province, China, are famous for their exquisite fossils that represent a diverse assemblage of invertebrates and vertebrates and include the preservation of both hard and soft tissues.

EXTENSIVE INLAND SEAS: THE MARINE PALEOZOIC

An extensive shallow sea may have covered much of the supercontinent itself. In these inland waters lived a community of organisms distinctively different to the Atdabanian fauna.

These waters were rich in brachiopods. Some burrowed into the sediments, while others (such as *Rafinesquina* and *Platystrophia*) made their home on the seabed and filtered food from the passing currents. There were trilobites of many kinds, including phacopids (such as *Flexicalymene*) and asaphids (such as *Isotelus*). Gastropods (such as *Cyclonema*) were also present, though less abundant and diverse than the brachiopods.

Thickets of crinoids (such as *Cincinnaticrinus*) colonized the seafloor alongside sponges, corals, and bryozoans, sometimes constructing reef-like environments. Corals (such as *Protaraea*) were encrusters, growing as thin sheets on the shells of other animals. Others grew singly, such as the horn corals *Streptelasma* and *Grewingkia*. Bryozoans also included encrusting forms, while other types (such as *Constellaria*) grew as branching colonies attached to the seafloor.

Lurking in the depths were large, predatory animals (such as the eurypterid *Megalograptus*). Other predators, such as the straight-shelled nautiloid *Isorthoceras*, cruised the waters above. And starfish (such as *Promopalaeaster*) crawled across the seabed in search of prey.

FOSSIL FAUNA OF THE CINCINNATIAN ROCKS OF OHIO

Fossils of the kinds described here can be found in great abundance in the Upper Ordovician rocks exposed along the interstate highways of the Cincinnati area of Ohio, USA. In fact, the trilobite *Isotelus maximus* is the state fossil of Ohio.

Road cuts near Cincinnati expose stacks of alternating thin layers of limestone and shale. The limestone beds are packed full of fossil brachiopods, bryozoans, and crinoids, often in a broken condition.

Only rarely are whole fossils found. It seems that during the Flood turbulent waves uprooted these creatures from the places where they were living and transported them some distance before burial.

▲ Upper Ordovician shales and limestones in a road cut along Interstate 75 in northern Kentucky, with the city of Cincinnati, Ohio, USA, in the background.

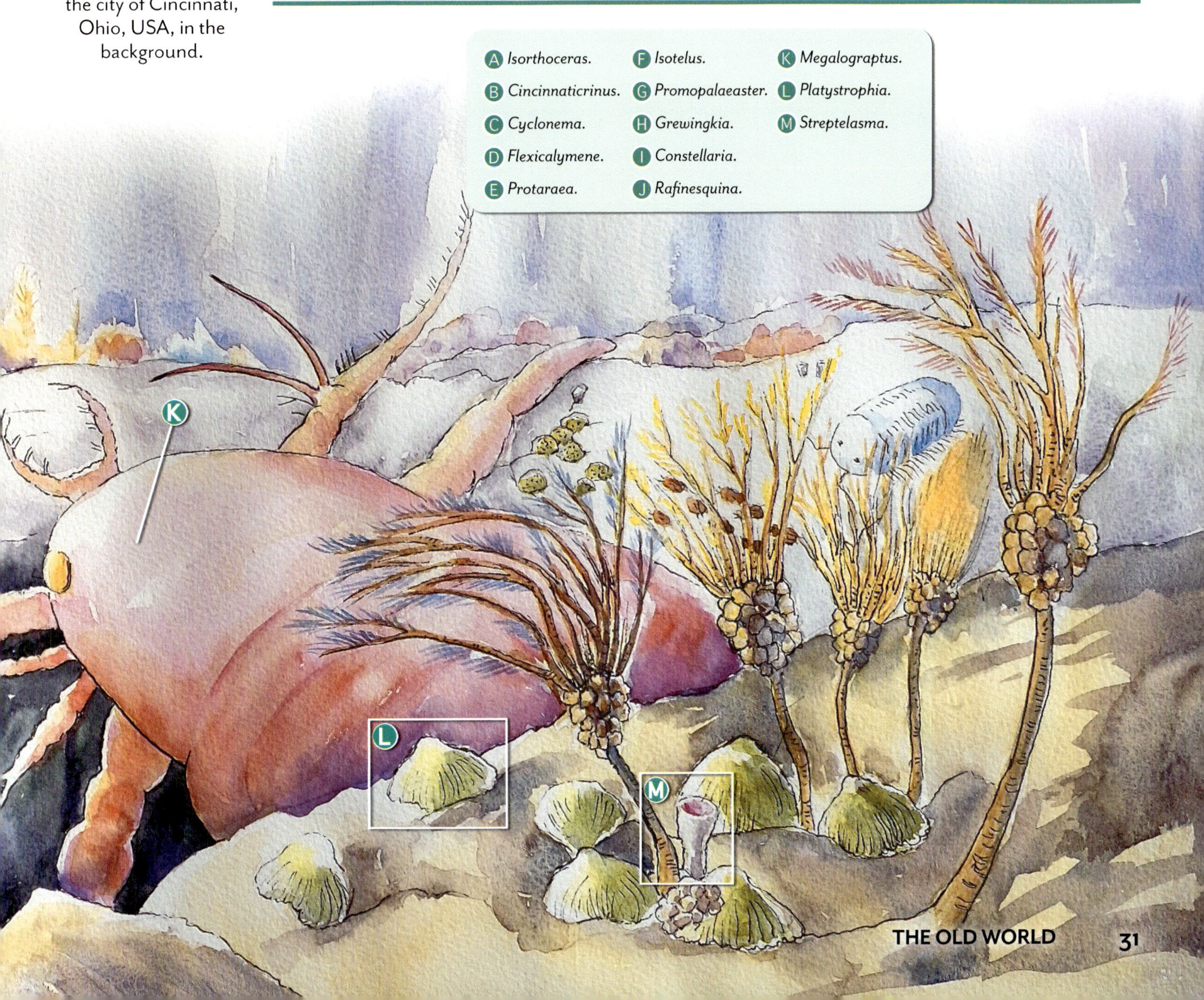

THE FRINGES OF THE LAND: COASTAL DUNES AND FORESTS

Reptiles thrived in the sand dunes and forests fringing the coasts of the pre-Flood supercontinent. Gymnosperm plants such as seed ferns and conifers dominated the forests.

Most of the animals inhabiting this biome were therapsids, a very diverse group with a peculiar mixture of reptile-like and mammal-like characteristics.

Most abundant were the dicynodonts—herbivores such as *Endothiodon* with two canines and a horny beak that they used for slicing and crushing plants.

There were also saber-toothed carnivorous therapsids called gorgonopsians. A typical example was *Lycaenops*, with its heavy skull, wide mouth, and massive canines.

Also common were the cynodonts, a group that included small- to medium-sized carnivores such as *Procynosuchus*.

Other therapsids included the dinocephalians, a group comprising bulky herbivores like *Moschops* and dog-sized carnivores like *Titanosuchus*.

Diapsid reptiles were also present, including the lizard-like *Youngina* and gliders such as *Coelurosauravus*.

FOSSIL REPTILES OF THE KAROO BASIN, SOUTH AFRICA

Large numbers of fossilized reptiles representing this extinct biome have been discovered in the Upper Permian sandstones and mudstones of the Karoo Basin in South Africa.

Robert Broom (1866-1951), a Scotsman who moved to South Africa at the turn of the twentieth century to practice medicine, made the largest collections. During his fifty-year scientific career, Broom named hundreds of fossil therapsids—including well-known types such as *Moschops* (in 1911), *Lycaenops* (in 1925), and *Procynosuchus* (in 1937)—as well as other reptiles such as *Youngina* (in 1914).

► Robert Broom, the paleontologist who discovered many of the fossil reptiles of the Karoo Basin, South Africa.

◄ The magnificent rock exposures in the Karoo desert are a rich source of vertebrate fossils.

DINOSAURS I. TRIASSIC BIOME

Large regions of the pre-Flood world must have been dominated by the extraordinary animals we call dinosaurs. In fact, it seems that there were at least three dinosaur biomes.

One of these biomes comprised the dinosaurs and associated creatures we find in Triassic rocks. Non-flowering gymnosperms dominated the flora, including conifers (such as *Protojuniperoxylon*). There were also horsetails and ferns (such as *Cladophlebis*).

The dinosaurs of the Triassic biome were mostly small, slender animals, including carnivores (such as *Herrerasaurus* and *Eodromaeus*) and plant-eaters (such as *Pisanosaurus* and *Chromogisaurus*).

But other kinds of reptiles were more numerous in this biome than the dinosaurs. There were the rhynchosaurs with their triangular skulls and sharp beaks; the aetosaurs with armored bodies, small heads, and upturned snouts; and the loricatans, crocodile-like in appearance but walking upright on four, long legs.

Therapsids were also represented by small forms (such as *Probainognathus*), medium-sized forms (such as *Chiniquodon*), and large, bulky forms (such as *Ischigualastia*).

DINOSAURS AND OTHER CREATURES OF ISCHIGUALASTO, ARGENTINA

Fossils representing this Triassic biome are preserved in the Ischigualasto Formation of the Valley of the Moon in northwestern Argentina.

Sparse dinosaur remains were discovered in this region in the 1950s, but our knowledge of these animals was greatly expanded when collecting resumed in the 1990s. The Ischigualasto dinosaurs include representatives of both main dinosaur subgroups: the ornithischians (bird-hips) and saurischians (lizard-hips).

Only about 10 percent of the vertebrate fossils in these rocks are dinosaurs. There are also fossils of many other reptiles, with the medium-sized rhynchosaur *Scaphonyx* accounting for more than half of the tetrapod remains in this formation.

▼ Wind-sculpted rock formations in the Valle de la Luna (Valley of the Moon) in Ischigualasto Provincial Park, San Juan Province, northwestern Argentina. These rocks belong to the Ischigualasto Formation, which has yielded many fossils of Triassic dinosaurs and other reptiles.

DINOSAURS II. JURASSIC BIOME

Another biome hosted the dinosaurs and associated animals and plants that we find in Jurassic rocks.

Conifers were the dominant trees in this biome, growing alongside mid-size plants such as ginkgos, cycads, tree ferns, and clubmosses. Small ground plants included horsetails and ferns (such as those belonging to the order Marattiales).

Giant sauropods like *Brachiosaurus* and *Apatosaurus* browsed the leaves from the tops of the tallest trees, while smaller ornithopods such as *Camptosaurus* and *Dryosaurus* browsed on the lower stems. Armored dinosaurs such as *Mymoorapelta* and *Stegosaurus* grazed on the small shrubs and ground plants.

The largest meat-eater in this biome was the theropod *Allosaurus*, but there were also smaller carnivores such as *Ceratosaurus*. The small, bird-like theropod *Ornitholestes* may have preyed on lizards and insects.

▶ The famous fossil wall at Dinosaur National Monument in Colorado, USA. About 1,500 dinosaur bones can be seen in the wall, which is part of the Jurassic Morrison Formation. The bones belong to *Allosaurus*, *Apatosaurus*, and *Stegosaurus*, among others.

Flying reptiles dominated the skies, including pterosaurs such as *Harpactognathus*. Scurrying around the feet of the dinosaurs were also small mammals. Some were burrowing animals, while others climbed trees. They all belonged to groups that are now extinct.

DINOSAURS AND OTHER CREATURES OF THE MORRISON FORMATION

The Morrison Formation is a widespread rock unit, extending for hundreds of miles across parts of seven American states: New Mexico, Oklahoma, Colorado, Utah, Montana, South Dakota, and Wyoming.

Hundreds of dinosaur skeletons have been excavated from these rocks since the days of the Great American Bone Wars, when Edward Drinker Cope (1840-1897) competed with Othniel Charles Marsh (1831-1899) for the best finds.

Many iconic dinosaurs are known from Morrison rocks, including meat-eaters such as *Allosaurus* and plant-eaters such as *Stegosaurus*. There are also fossils of many other animals, including lizards, turtles, crocodiles, mammals, fishes, and insects, as well as many diverse plants.

DINOSAURS III. CRETACEOUS BIOME

The dinosaurs and associated animals and plants that we find in Cretaceous rocks inhabited another biome.

Large trees included the fruit-bearing angiosperms such as sycamores and magnolias. Mid-size plants included laurels, and there were small ground plants such as the fern *Dryopteris*.

Browsing and grazing on these plants were herbivores such as the hadrosaur *Edmontosaurus*, the horned dinosaur *Triceratops*, the armored *Ankylosaurus*, and the bone-headed *Pachycephalosaurus*.

The largest meat-eater in this biome was *Tyrannosaurus*, but there were smaller carnivores such as *Dromaeosaurus* and *Troodon*, both of which probably possessed a feathery covering.

Pterosaurs included true giants such as *Quetzalcoatlus*, with an estimated wingspan of 33 to 36 feet (10 to 11 meters). This extraordinary pterosaur was one of the largest flying animals ever to have lived. Small birds such as *Avisaurus* and small mammals such as *Alphadon* and *Mesodma* also inhabited this biome.

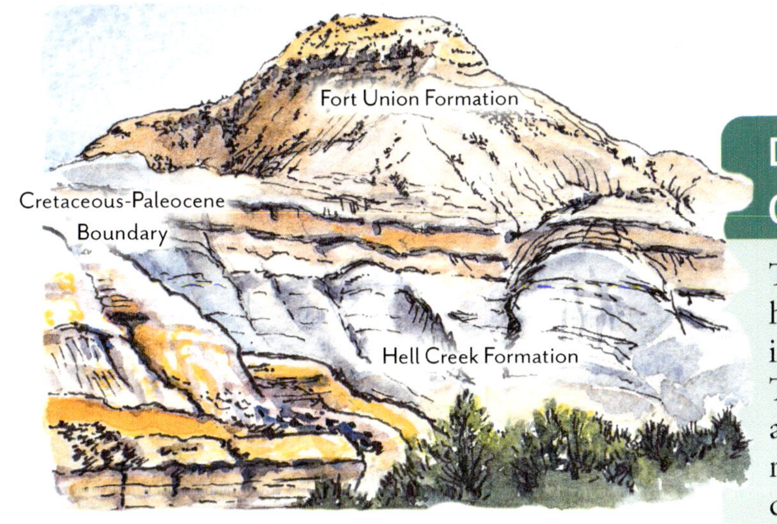

▲ The Cretaceous-Paleocene boundary in Makoshika State Park near Glendive, Montana, USA, is marked by a narrow band of iridium-enriched clay and carbonized plant material. The dinosaur-bearing Hell Creek Formation (Cretaceous) lies below the dark band; the Fort Union Formation (Paleocene) lying above it does not have dinosaur fossils.

DINOSAURS AND OTHER CREATURES OF THE HELL CREEK FORMATION

The Hell Creek Formation in Montana, USA, has yielded many fossil skeletons belonging to inhabitants of the Cretaceous dinosaur biome. These include dinosaurs and pterosaurs, but also crocodiles, lizards, turtles, frogs, fishes, and mammals. Among the most famous Hell Creek dinosaurs are the top predator, *Tyrannosaurus*, and the horned dinosaur, *Triceratops*.

Scientists have studied the Hell Creek Formation extensively because these rocks include the boundary that marks the last appearance of the dinosaurs in the fossil record. The thin clay layer at the boundary is enriched with the element iridium, which may have come from an asteroid that hit the earth when these rocks were being formed.

MARINE REPTILE BIOMES

The world before the Flood was also home to a greater diversity of marine reptiles than are living today. These animals may have inhabited the warm, shallow waters of inland seas closely associated with the dinosaur biomes of the Triassic, Jurassic, and Cretaceous.

Among the best known of these fossilized marine reptiles are the ichthyosaurs (meaning *fish lizards*). They had streamlined bodies with front and rear paddles, and a deep tail fin. Their long, thin snouts were equipped with an array of sharp teeth, and they seem to have lived on a diet of fish and shellfish.

Another well-known group is the plesiosaurs (meaning *ribbon lizards*). They had compact bodies and short tails but elongated necks. They probably used their paddles to swim with an underwater flying motion, darting out their long necks to catch fish and squid.

Perhaps most spectacular of all were the giant marine lizards called mosasaurs. These animals had long bodies, deep tails, and paddle-like limbs. Their wide jaws were lined with sharp, conical teeth, which they used to catch fish and crack open mollusk shells.

Other marine reptiles included the nothosaurs, slender animals with four paddle-like limbs, and bulky mollusk-eaters such as the placodonts, along with more familiar groups such as crocodiles and sea turtles.

The Triassic marine reptile biome was home to ichthyosaurs (such as *Mixosaurus*), placodonts (such as *Placodus*), and nothosaurs (such as *Lariosaurus*). Also shown is the ray-finned fish, *Saurichthys*. This reconstruction is based on fossils from the Middle Triassic "Lower Reptile Bed" of the Guanling Formation in Guizhou Province, China.

The Jurassic marine reptile biome was inhabited by ichthyosaurs (such as *Stenopterygius*), plesiosaurs (such as *Plesiosaurus*), and crocodiles (such as *Steneosaurus*). Ammonites (such as *Dactylioceras*) and ray-finned fishes (such as *Lepidotes*) also swarmed in these warm waters. This reconstruction is based on fossils from the Lower Jurassic Posidonia Shale of Holzmaden, Germany.

The Cretaceous marine reptile biome was populated by mosasaurs (such as *Tylosaurus*), plesiosaurs (such as *Elasmosaurus*), and sea turtles (such as *Archelon*). Also shown is *Squalicorax* (a shark), *Protosphyraena* (a swordfish), and *Pachyrhizodus* (a bony fish). This reconstruction is based on fossils from the Upper Cretaceous Niobrara Chalk of Kansas, USA.

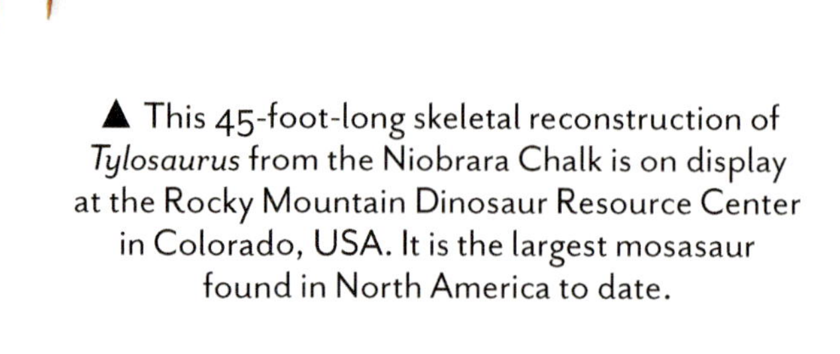

▲ This 45-foot-long skeletal reconstruction of *Tylosaurus* from the Niobrara Chalk is on display at the Rocky Mountain Dinosaur Resource Center in Colorado, USA. It is the largest mosasaur found in North America to date.

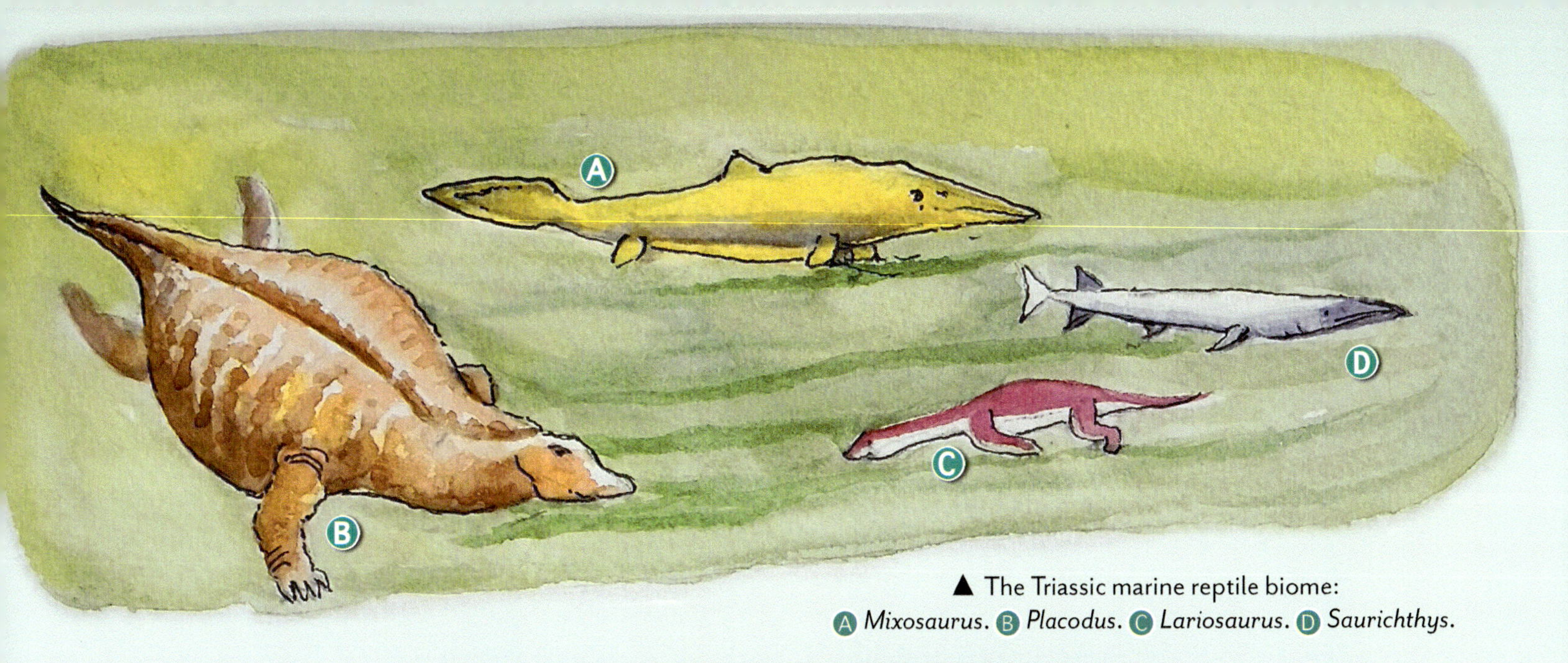

▲ The Triassic marine reptile biome:
Ⓐ *Mixosaurus*. Ⓑ *Placodus*. Ⓒ *Lariosaurus*. Ⓓ *Saurichthys*.

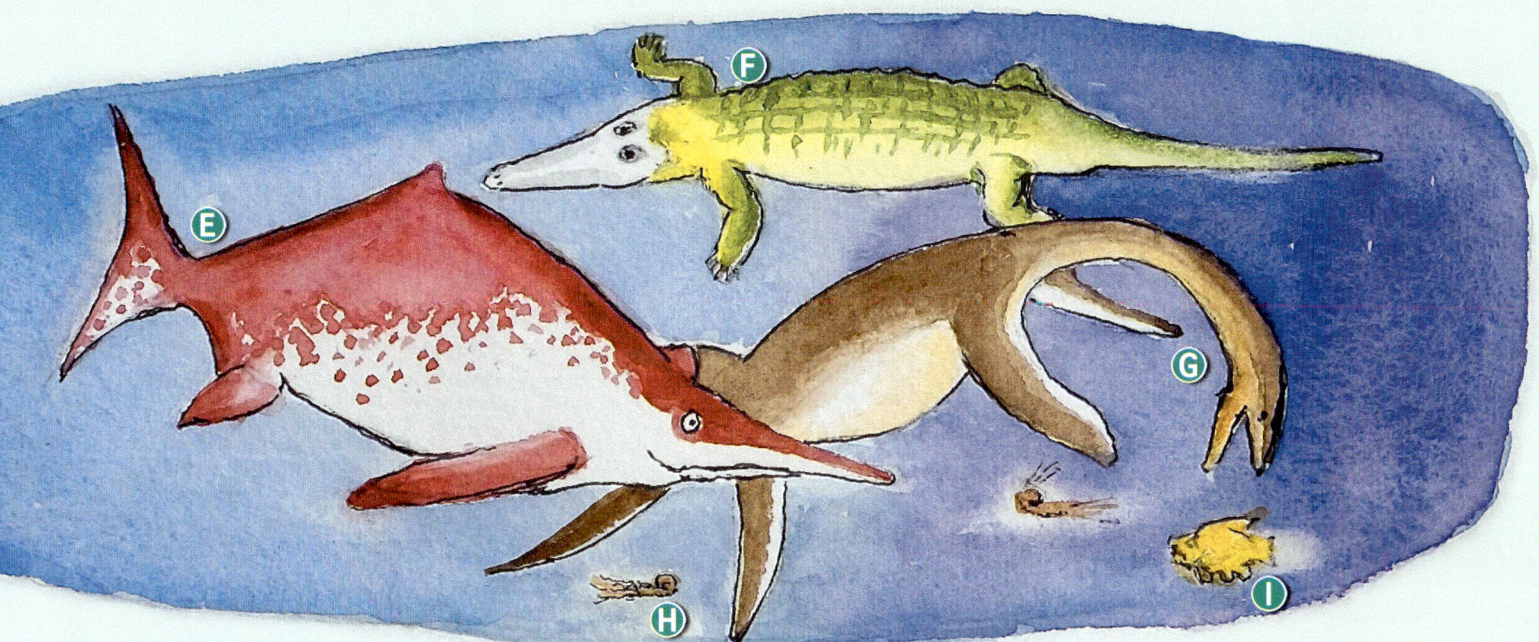

▲ The Jurassic marine reptile biome:
Ⓔ *Stenopterygius*. Ⓕ *Steneosaurus*. Ⓖ *Plesiosaurus*. Ⓗ *Dactylioceras*. Ⓘ *Lepidotes*.

▼ The Cretaceous marine reptile biome:
Ⓙ *Elasmosaurus*. Ⓚ *Squalicorax*. Ⓛ *Pachyrhizodus*. Ⓜ *Protosphyraena*. Ⓝ *Archelon*. Ⓞ *Tylosaurus*.

EDEN AND ITS SURROUNDINGS

There must have been at least one other biome before the Flood—one in which most of the mammals and birds lived alongside humans. The dominant plants in this biome were probably angiosperms (flowering plants).

For reasons we will explore later, this biome seems to be entirely missing from Flood sediments. This means that we must speculate to a considerable degree about what this biome and its inhabitants were like.

We can get some clues about this biome from the Bible's description of the Garden of Eden and its surroundings. Genesis 2:10-14 tells us that a river ran out of Eden and split into four rivers that ran into four different countries. This suggests that Eden was at a higher elevation than the lands surrounding it.

Making the reasonable assumption that the humans were living in a biome centered around Eden, it seems likely that this was an upland ecosystem.

Furthermore, the river in Eden may have been fed by a spring bringing water from below the ground. Genesis 2:6 refers to a mist that came

up to water the ground. Although the precise nature of this mist is not clear, the use of the same word in languages similar to biblical Hebrew suggests that it may have been a flowing spring of some kind.

Perhaps this biome was located right on top of one of the "fountains of the great deep" that broke open at the beginning of the Flood (Genesis 7:11).

We can also infer that humans living in this biome had an advanced culture. The Bible tells us that before the Flood, humans were farming arable crops and livestock (Genesis 4:2), building cities (Genesis 4:17), composing music, and working with metals (Genesis 4:19-22).

The next section of the book will look at the destruction of these biomes in the worldwide catastrophe of Noah's Flood.

This biome and its inhabitants were not preserved in Flood rocks, so we have reconstructed them here based on clues from the Bible and the earliest post-Flood fossil record:

- A Ape (based on *Aegyptopithecus* from the Eocene-Oligocene).
- B Lemur (based on *Plesiadapis* from the Eocene).
- C Hawk (based on *Milvoides* from the Eocene).
- D Mesonychid (based on *Dissacus* from the Paleocene-Eocene).
- E Kingfisher (based on *Primobucco* from the Eocene).
- F Heron (based on *Zeltornis* from the Miocene).
- G Raoellid (based on *Indohyus* from the Eocene).
- H Pantolestid (based on *Buxolestes* from the Eocene).
- I Chalicothere (based on *Schizotherium* from the Eocene-Oligocene).
- J Humans (based on early *Homo* from the Plio-Pleistocene).
- K Songbird (based on *Resoviaornis* from the Oligocene).
- L Dog (based on *Hesperocyon* from the Eocene-Oligocene).

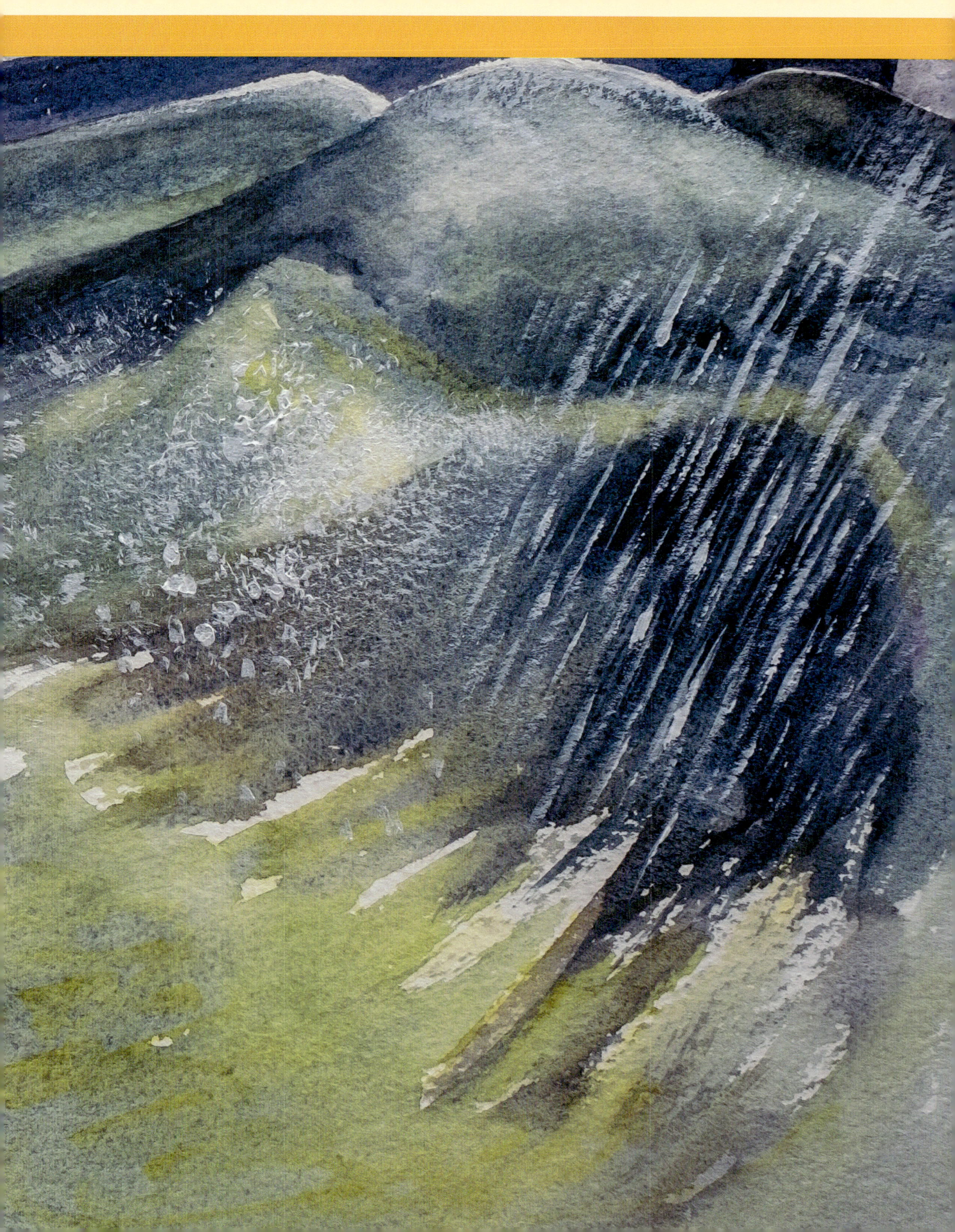

3

THE OLD WORLD DESTROYED

Noah's Flood was a judgement on human sin that destroyed the old world. During the Flood, the continents were broken apart and rearranged, causing the oceans to inundate the land. Whole communities of plants and animals were ripped up from where they had been living and were transported, before being rapidly buried in layers of water-deposited sediment spanning the continents. At the end of the Flood, the waters drained from the land and returned to the deepening ocean basins.

THE FLOOD ANTICIPATED

The downward spiral of sin that led to the Flood began when the first man, Adam, rebelled against God in the Garden of Eden (Genesis 3:1-6). His sinfulness was passed on to all his descendants, bringing death and bloodshed in its wake (Genesis 3:17-19; Romans 5:12).

The next generation witnessed the first murder, when Adam's eldest son, Cain, slew his brother, Abel, in a fit of jealous rage (Genesis 4:1-8). Soon others were seeking vengeance (Genesis 4:23-24). As the people multiplied in number, the whole earth became filled with violence (Genesis 6:11).

Eventually, the time came when God looked upon the world that he had made and saw that all flesh—both men and animals—was corrupted (Genesis 6:5-7; see also 6:12-13 and 7:21-23):

> *And God saw that the wickedness of man was great in the earth, and that every imagination of the thoughts of his heart was only evil continually. And it repented the LORD that he had made man on the earth, and it grieved him at his heart. And the LORD said, I will destroy man whom I have created from the face of the earth: both man, and beast, and the creeping thing, and the fowls of the air: for it repenteth me that I have made them.*

Yet even as God determined to bring the judgement of the Flood upon the world, he was also ready to display his grace. One man, Noah, found favor in God's sight (Genesis 6:8). Warning Noah of the impending Flood, God instructed him to build an ark in which he, his family, and representatives of all the birds and air-breathing land animals might be saved (Genesis 6:14-21).

Noah obeyed God and built the ark according to the instructions he had been given (Genesis 6:22). The completed ark was 300 cubits long, 50 cubits wide, and 30 cubits high. Assuming a long cubit of about 20.4 inches (51.8 centimeters), the ark could have measured about 510 feet (155 meters) long, 85 feet (26 meters) wide, and 51 feet (16 meters) high. It included a window, a door in the side, and three decks.

Noah gathered the birds and the animals that God brought to him, and he took into the ark food of every kind for himself and the animals (Genesis 6:21). Eventually the day came when God told him to go into the ark with his family and all the animals (Genesis 7:1-5). Then God himself shut them in (Genesis 7:16).

WAS THE FLOOD TRULY GLOBAL?

There are many reasons to think that Noah's Flood covered the whole earth, and not just a local area or region.

1. The purpose of the Flood was to wipe mankind "from the face of the earth" (Genesis 6:7)—and we are told that the only human survivors were Noah and his family (1 Peter 3:20; 2 Peter 2:5).
2. No local flood could cover the tops of the highest mountains or prevail for ten months as the Bible says Noah's Flood did (Genesis 7:19-20; 8:5).
3. There was no reason for Noah to build the ark if the Flood were only local in extent—he could have migrated out of the area or to higher ground.
4. The water was on "the face of the whole earth" (Genesis 8:9)—a phrase used four times in the book of Genesis outside the Flood account and always in the universal sense.
5. All the high hills "under the whole heaven" were covered (Genesis 7:19)—another phrase that is used in a universal sense throughout the Old Testament (except in one case where it is qualified).
6. The covenant that God made with Noah and his descendants (Genesis 9:8-9) cannot be applied to the whole of humanity if the destruction of the Flood was not universal.
7. God promised that he would never send another flood like this one (Genesis 9:11-16)—but God has broken his promise repeatedly if Noah's Flood were only local.
8. The Lord Jesus, Peter, and the writer to the Hebrews spoke of the Flood in universal terms as a picture of the judgement to come (Luke 17:27; Hebrews 11:7; 1 Peter 3:20; 2 Peter 2:5; 3:6). Peter portrays the Flood as an event that dismantled the original creation (2 Peter 3:4-6), confirming its worldwide extent.

HOW MANY ANIMALS WERE ON THE ARK?

Noah was commanded to take on board the ark the flying creatures and the air-breathing land animals. These would have been mostly mammals, birds, and reptiles. Aquatic animals were able to survive outside the ark. It is probable that many of the invertebrates (such as insects) were not taken on board either.

Furthermore, Noah did not have to care for every *species*, but rather representatives of each *kind*. The biblical kind is much broader and more inclusive than the species, and it probably approximates the family in modern biological classifications.

Recent efforts to tally the ark kinds for the Ark Encounter in Kentucky, USA, based on numbers of families suggest that there may have been 468 mammal kinds, 320 reptile kinds, 78 mammal-like reptile kinds, 284 bird kinds, and 248 amphibian kinds. This gives a total of 1,398 ark kinds.

Genesis 7:8-9 says that the animals came into the ark by twos, and the clean animals and the flying creatures by sevens. The Hebrew wording with respect to the clean animals may mean seven pairs or seven individuals. Assuming the higher figure, there would have been around 6,744 animals—a number easily accommodated on a vessel the size of the ark.

THE FLOOD BEGINS

Genesis 7:11 tells us how the Flood began:

> *In the six hundredth year of Noah's life, in the second month, the seventeenth day of the month, the same day were all the fountains of the great deep broken up, and the windows of heaven were opened.*

The fountains of the great deep seem to have been water sources distributed across the earth's surface, both on the continents and in the oceans. Noah's Flood began when all these fountains were broken up on a single day.

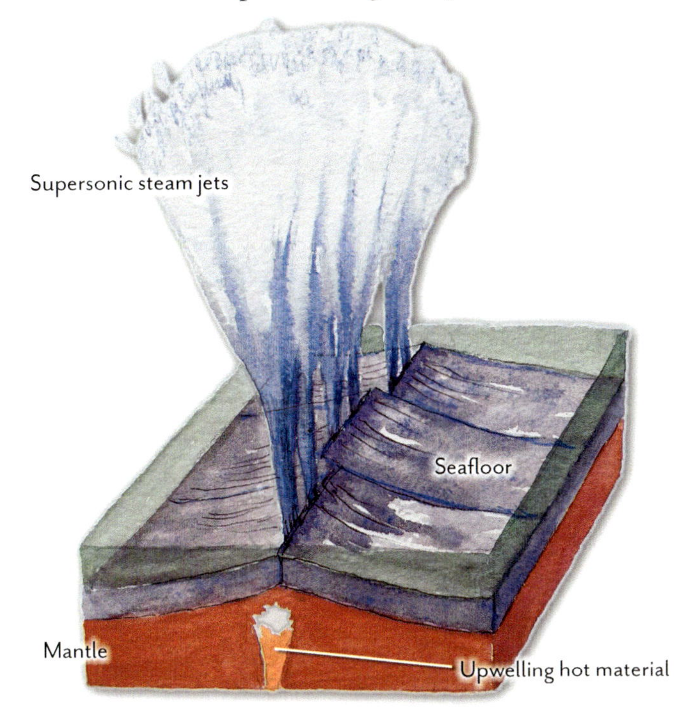

The phrase "broken up" is used several times in the Old Testament to describe faulting or cleaving of the earth's surface (see Numbers 16:31; Judges 15:19; Psalm 78:15; Isaiah 48:21; Micah 1:4; Zechariah 14:4).

The implication seems to be that the Flood began with some kind of upheaval involving the breaking up of the earth's crust, leading to fountains of water being released from beneath the surface of the land and the oceans.

MEGAQUAKES AT THE START OF THE FLOOD

Geological evidence indicates that massive earthquakes many times larger than any in the modern world caused the margins of the pre-Flood supercontinent to collapse at the beginning of the Flood.

A powerful testimony to these catastrophic events can be seen in the rocks of the Kingston Peak Formation of Death Valley, California, USA. Embedded within these sedimentary layers are huge blocks of rock, some hundreds of meters long. They seem to represent the broken fragments of a collapsed ocean shelf.

At the beginning of the Flood, enormous underwater avalanches carried these blocks downslope, where they were buried by other sediments. Similar rocks at the same level of the **geological column** are found in other places around the world.

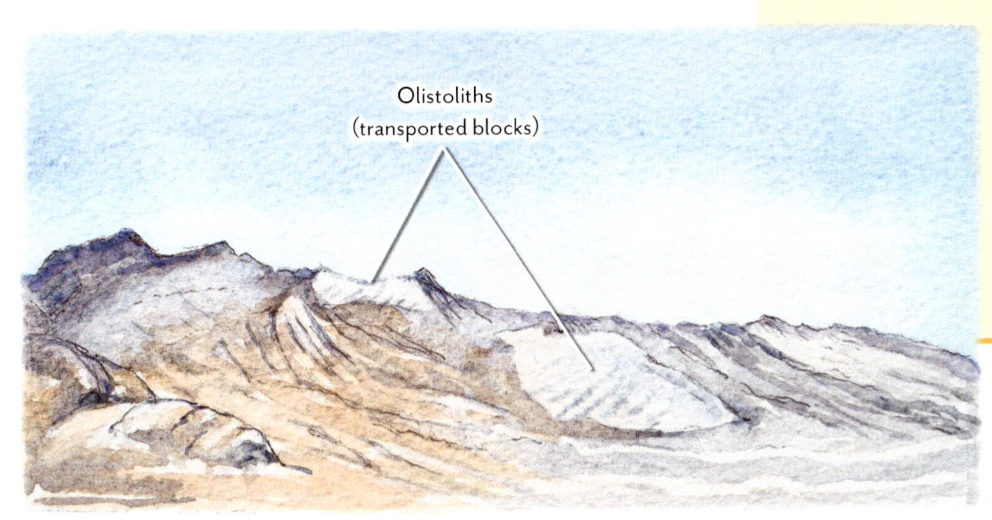

◀ Part of the Precambrian Kingston Peak Formation in the Kingston Range of Death Valley, California, USA. Highlighted are two of the enormous blocks of rock embedded within the formation. The largest is about 1,640 feet (500 meters) long.

THE GREAT UNCONFORMITY

As the sediment-laden floodwaters rose and swept across the land, the continents were ground down to basement level—even down to the crystalline rocks that make up the stable core of the continents. Flood sediments began to build up on top of these eroded basement rocks, leaving what has come to be known as the Great Unconformity.

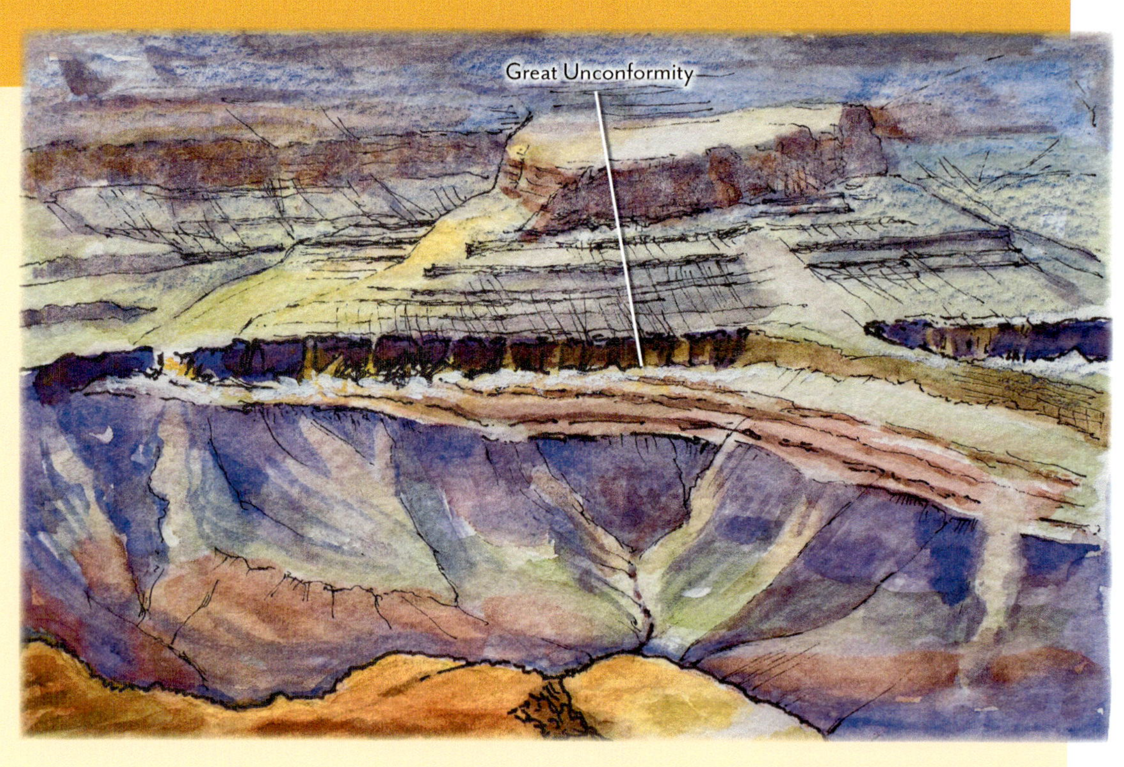

▲The Great Unconformity from the Desert Overlook on the southeast rim of the Grand Canyon, Arizona, USA. A sequence of horizontally bedded sandstones, shales, and limestones overlies the tilted rocks of the Precambrian Grand Canyon Supergroup. At this locality, the boundary between these two sets of rocks marks the beginning of the Flood.

The Great Unconformity is an erosion surface that separates older, pre-Flood basement rocks from **sedimentary rocks** on top that were deposited during the Flood. It is a truly remarkable geological feature that can be observed on at least five continents (North America, Europe, Asia, Africa, and Antarctica).

It also marks the first appearance in the fossil record of the diverse Cambrian fossils—multicellular animals with hard, external skeletons. These were among the earliest animals to be buried during the Flood.

THE FLOOD UNFOLDS

We know today that the earth has a layered structure consisting of a **core**, **mantle**, and **crust**. The crust is the thin outer layer and is broken into a series of rigid plates that are able to move relative to one another.

Insights from modern geology combined with clues from the Bible suggest that these **tectonic plates** were broken apart at the time of Noah's Flood.

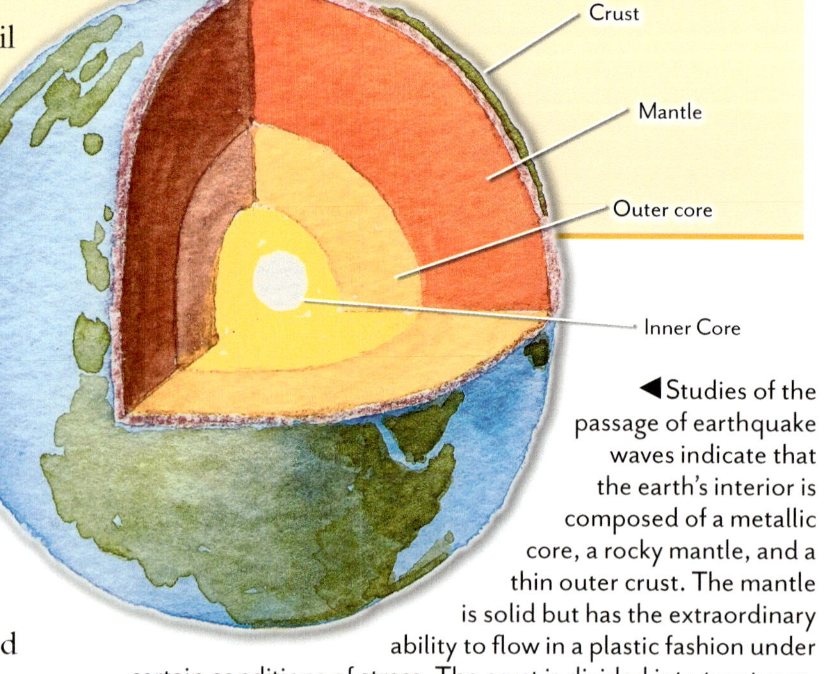

◀Studies of the passage of earthquake waves indicate that the earth's interior is composed of a metallic core, a rocky mantle, and a thin outer crust. The mantle is solid but has the extraordinary ability to flow in a plastic fashion under certain conditions of stress. The crust is divided into two types. Oceanic crust is made of a dense rock called basalt, which sits lower in the earth's mantle and forms the ocean basins. Continental crust is composed of a lighter rock called granite, which floats higher in the mantle and forms the world's landmasses.

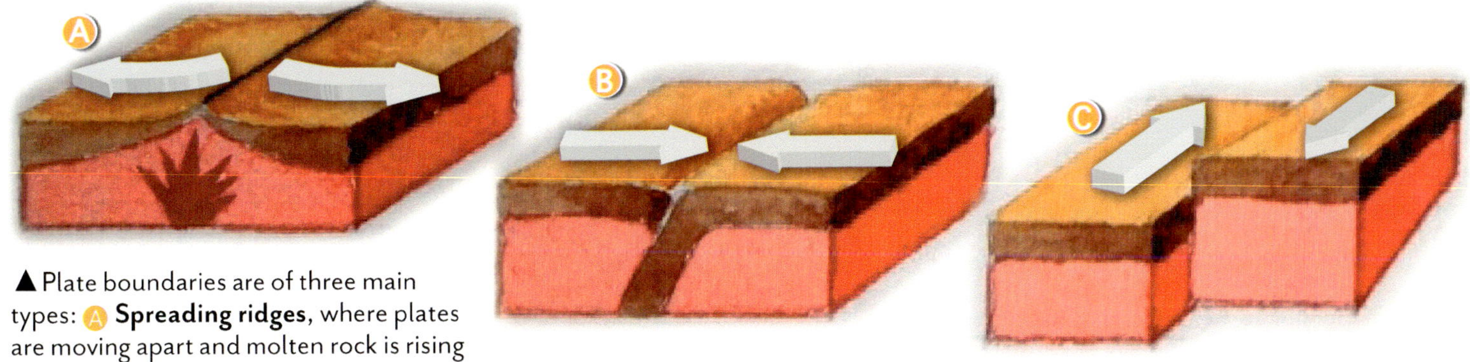

▲ Plate boundaries are of three main types: (A) **Spreading ridges**, where plates are moving apart and molten rock is rising up to add new ocean floor material between. (B) **Subduction zones**, where plates are moving towards one another and one plate is sinking under the other and being consumed. (C) **Transform faults**, where two adjacent plates are slipping past one another without the addition or destruction of plate material.

During the Flood, three things happened at the same time:

1. The pre-Flood ocean crust broke loose from the margins of the supercontinent and began to plunge into the earth's mantle, pulling the plates apart and causing the break-up of the pre-Flood landmass.
2. The diving ocean floor sank rapidly through the mantle, pushing the surrounding hot material out of the way and stirring the mantle throughout its entire depth.
3. Hot mantle material welled up to form new ocean crust along the mid-ocean ridges where the plates were separating.

Since the new ocean crust was warmer (and, therefore, less dense) than the old ocean crust it was replacing, it was more buoyant and rode higher in the earth's mantle. As a consequence, the level of the ocean floor rose significantly, causing the ocean basins to become shallower and displacing water onto the continents.

Sea level may have risen more than 1 mile (1.6 kilometers) over its pre-Flood level, enough to flood the continents and cover the tops of even the highest pre-Flood mountains.

Moreover, along the mid-ocean ridges where new hot material was coming into contact with cold ocean water, the water was vaporized and propelled high into the atmosphere. As a result, supersonic steam jets erupted along thousands of kilometers of mid-ocean ridge. These geysers were probably among the fountains of the great deep mentioned in Genesis 7:11.

The ocean water that was caught up by these supersonic jets fell back to the earth as an intense, global rain. This may have been the primary source of the rain that fell from the "windows of heaven" for forty days and nights (Genesis 7:4, 11).

And all the while, as the world was underwater, catastrophic plate movements were rapidly rearranging the shattered fragments of the supercontinent—breaking them apart and crashing them back together before finally separating them again to produce the modern-day continents.

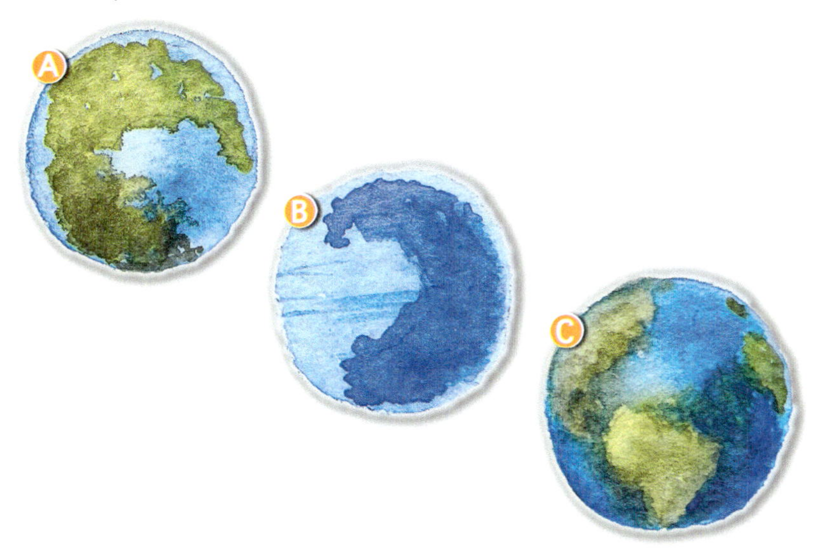

▲ (A) The original pre-Flood supercontinent Rodinia broke apart early in the Flood. (B) In the middle of the Flood, the fragments briefly came together again to form another supercontinent called **Pangea**, but this supercontinent was submerged beneath the floodwaters. (C) Later in the Flood, Pangea broke apart to produce the familiar continents and oceans of today's world.

RUNAWAY SUBDUCTION

Plates diving into the earth's interior today pass through the mantle very slowly because the mantle material is resistant to flow.

Geophysicist John Baumgardner has proposed that under certain conditions, plates can move through the mantle at much faster rates than at present. The Flood would have provided these conditions.

Friction from a plunging slab causes the mantle around the slab to heat up. As a result, the mantle's resistance to flow decreases and the plunging slab accelerates. As the speed of the slab increases, additional frictional heating takes place, causing around the slab a further decrease in resistance to flow.

During the Flood, sinking slabs accelerated until they were sinking at rates of meters per second—billions of times faster than observed in the present day.

▼Slabs of ocean crust sinking through the mantle generate frictional heating, which in turn leads to a weakening of the surrounding mantle rocks. If conditions are right, this weakening can lead to an increased sinking rate, an increased heating rate, and even greater weakening. This positive feedback can result in thermal runaway, or **runaway subduction**.

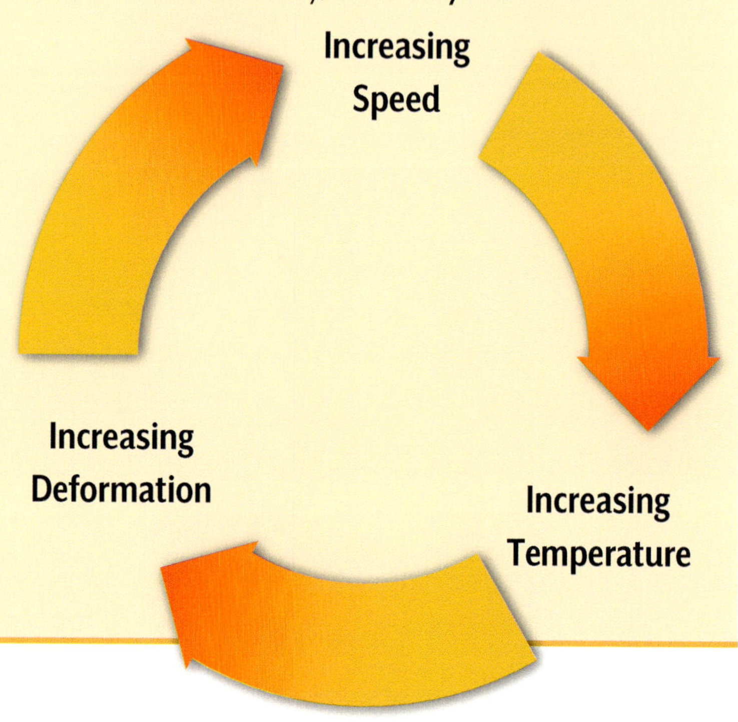

BIOMES ARE BURIED

As the floodwaters advanced onto the pre-Flood supercontinent, communities of organisms were picked up, transported, and buried in the order in which they were encountered. In this manner, a horizontal series of pre-Flood biomes was buried during the Flood to produce a vertical sequence of rock layers with characteristic fossils.

▼The sequential burial of a horizontal series of pre-Flood biomes produced a vertical sequence of rock layers with characteristic fossils.

First the biomes on the continental shelf were buried: ① the stromatolite reef, ② the Ediacarans, ③ the small shelly animals and ④ the Atdabanian animals ...

⑤ Next, the marine Paleozoic animals were buried ...

⑥ ... and while that was happening, the floating forest was drifting across the submerged continents and being ripped apart.

⑦ Then the coastal dunes and forests were buried ...

First to be overwhelmed were the stromatolite reefs that formed a barrier around parts of Rodinia, separating the deep ocean from the shallow marine shelf. This barrier was breached when the continental margins collapsed at the beginning of the Flood. Then the rising floodwaters poured in, overwhelming the animals living in the marine lagoon behind the reef: first the Ediacarans, then the small shelly creatures, and then the Atdabanian animals. The burial of the Lower Cambrian Atdabanian fauna accounts for the sudden appearance in the fossil record of abundant and diverse hard-shelled animals—an event that geologists have called the Cambrian explosion.

Next, the animals populating the shallow seas that extended over the supercontinent began to be buried. These are the animals of the marine Paleozoic. While this was happening, the floating forest drifted in over the submerged continents, perhaps because plate movements were causing continental fragments on either side of the floating forest to move towards one another. Tsunamis breaking over the submerged continents disrupted the floating forest, ripping it up and burying the plants from the outside in.

First, the small, water-dependent plants with rhizomes were buried, followed by the larger plants with true roots that were less dependent on water. Insects and animals living in the floating forest were overwhelmed too: the fishes and tetrapods living in pools on the forest surface, as well as the animals inhabiting the drier, more central parts of the floating forest. One of these was the reptile *Hylonomus*, whose fossilized remains are sometimes found inside the hollow trunks of lycopsid trees, where the animals were apparently taking refuge when they were buried.

By the time nearly the entire Paleozoic marine fauna had been buried, the floating forest had been completely destroyed. Only the logs were left to produce the coal layers of the Carboniferous.

This scenario helps explain why plants from the floating forest are commonly found in layers of sediment sandwiched between those containing Paleozoic marine fossils, especially in Silurian and Devonian rock sequences.

Next, the floodwaters began to encroach upon the land. First, the animals and plants living around the coasts were overwhelmed—mostly therapsid reptiles and gymnosperms. Then the floodwaters reached the dinosaurs—first those in the Triassic biome, then those in the Jurassic biome, and finally those in the Cretaceous biome. It is likely that these biomes were adjacent to one another, perhaps each successively farther inland.

Finally, humans, mammals, and birds were overwhelmed. Since no fossils representing this biome have been found in sediments deposited during the Flood, it is possible that this biome

8 Followed by the Triassic dinosaurs and marine reptiles ...

9 The Jurassic dinosaurs and marine reptiles ...

10 The Cretaceous dinosaurs and marine reptiles ...

11 And finally the humans, and most mammals and birds.

was completely destroyed—especially if it was located on top of one of the fountains of the great deep or alongside a subduction zone. If it was not completely destroyed, it was probably the last to be buried since it was located in an upland region. It may have been the first to be washed away by the floodwaters when they drained off the land at the end of the Flood.

DID YOU KNOW?

The idea that the order of the fossil record reflects the order in which different ecosystems were buried during the Flood is known as the **ecological zonation theory**. Creationist Harold Willard Clark first put it forward in 1946 in a book called *The New Diluvialism*. Other creationists have since developed and modified it.

TRACKWAYS BEFORE BODY FOSSILS

A very curious feature of the geological record is the tendency for the fossilized trackways of amphibians and reptiles to occur in sedimentary layers below the bones of the animals that made them.

For example, the first dinosaur trackways occur in rock layers below those containing the first dinosaur body fossils. And, overall, dinosaur tracks are most abundant and diverse in Triassic and Lower Jurassic rocks, while dinosaur body fossils are most diverse in Cretaceous rocks.

Something similar is evident in the invertebrate track record. In every location where the lowest rock layers containing trilobite fossils are found, trilobite tracks (called *Cruziana*) occur below the first trilobite body fossils.

This pattern seems to be best explained by the Flood model. As sediments were being laid down during the Flood, animals were initially able to move around (leaving footprints or tracks) only to succumb later (leaving their skeletal remains).

EVIDENCE OF RAPID BURIAL

During the Flood, sediments built up quickly and buried enormous numbers of plants and animals. Many fossils show clear evidence of rapid burial associated with the Flood.

Fishes are often found in large numbers, sometimes with delicate soft tissues preserved. Occasionally specimens were fossilized while eating another fish. The Lower Cretaceous Santana Formation of Brazil contains fishes whose gills and muscles are so perfectly preserved that geologists believe they were fossilized within five hours of death. In fact, it has been suggested that most of the Santana Formation fishes died because they were being fossilized as they swam!

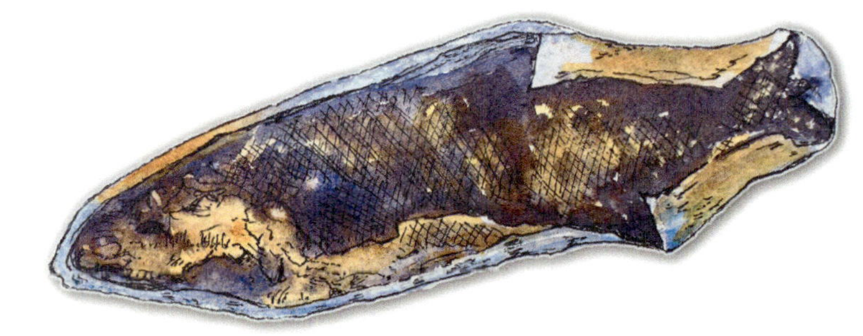

▲ Fossilized fish in a carbonate concretion from the Santana Formation of northeastern Brazil. Some of these specimens preserve exquisite details of soft parts, including gills, muscles, and stomachs.

▲ During the Flood, trilobites may have burrowed their way upwards through the accumulating sediment, leaving their tracks on each successive layer, until finally they succumbed from exhaustion. This may explain why we find trilobite tracks below trilobite body fossils in the lowest trilobite-bearing layers.

The Jurassic sediments at Holzmaden, Germany, are famous for their complete skeletons of sharks, fishes, and marine reptiles such as ichthyosaurs. On many of these skeletons, the skin is preserved as a black carbon film. Hundreds of specimens of the ichthyosaur *Stenopterygius* have been found here, sometimes with stomach and intestine contents. There are even female ichthyosaurs fossilized while giving birth to young.

Fossils of wholly soft-bodied organisms are rare, but jellyfish fossils have been found in Cambrian sediments in Utah and Wisconsin, USA, among other places. The remarkable preservation of these fossils is the result of rapid burial, which halted the normal processes of scavenging and decay.

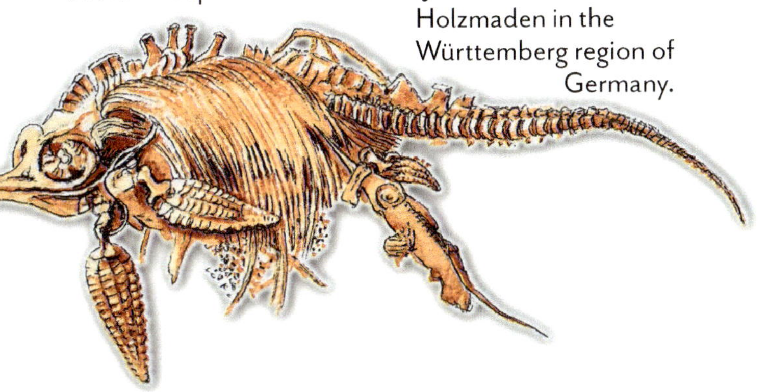

▼An ichthyosaur captured as a fossil while giving birth. This specimen came from Jurassic sediments near Holzmaden in the Württemberg region of Germany.

◀An enrolled trilobite (*Flexicalymene*) from the Ordovician of Ontario, Canada.

►An exceptionally well-preserved jellyfish fossil from the Marjum Formation (Middle Cambrian) of Utah, USA. Specimens in these rocks reveal details of the soft parts including trailing tentacles.

Other fossil vertebrates—including dinosaurs—are often found in a peculiar pose, with the mouth gaping, the head thrown back, and the tail arched. Experts have long debated the causes of this strange posture. Recent research suggests that it may have resulted from the death throes typical of asphyxiation (oxygen deprivation) or the immersion of carcasses in water. The commonness of this posture in fossilized vertebrates is consistent with drowning as the cause of death.

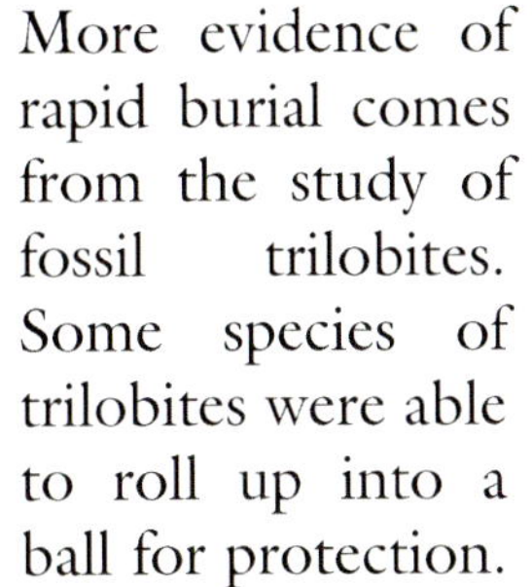

More evidence of rapid burial comes from the study of fossil trilobites. Some species of trilobites were able to roll up into a ball for protection. Many trilobites are found fossilized in this position, indicating that these animals were buried alive while trying to protect themselves.

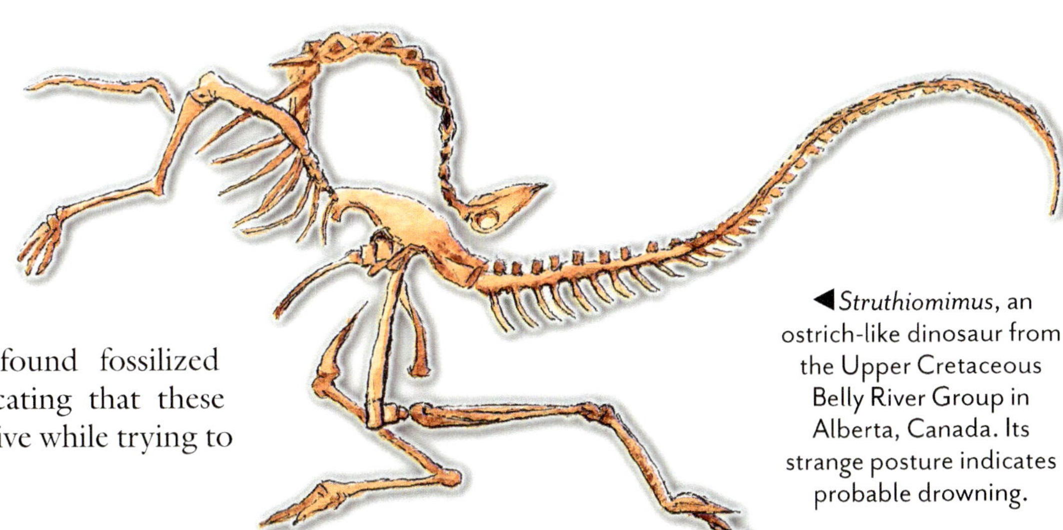

◀*Struthiomimus*, an ostrich-like dinosaur from the Upper Cretaceous Belly River Group in Alberta, Canada. Its strange posture indicates probable drowning.

In modern environments, even hard parts such as shells will eventually break down or dissolve in seawater, and fragile shells will break down more quickly than strong shells. If the fossil record formed slowly, with individual rock layers taking hundreds or thousands of years to accumulate, we would expect fragile shell material to be relatively uncommon. Most of the shells in the fossil record should be thick and durable. But studies show that small, thin-shelled fossils are just as likely to be preserved in the record as large, thick-shelled fossils. This is further evidence that the sediments in which they are buried were formed rapidly.

Finally, there are countless examples of tree trunks fossilized in upright positions and sometimes penetrating multiple sedimentary layers. They are especially common in coal-bearing sediments, such as those exposed in the cliffs along the Bay of Fundy in Nova Scotia, Canada. Geologists explain the preservation of these trees by rapid burial before the normal processes of decomposition could take place. Often, as in the case of the petrified forests of Yellowstone National Park in Wyoming, USA, there is evidence that the fossilized trees are not in the place where they originally grew but were uprooted and transported into place before burial.

▼The upright trunks of three lycopsids in Carboniferous sediments exposed along the shoreline near the town of Joggins in Nova Scotia, Canada.

HOW COMPLETE IS THE FOSSIL RECORD?

The standard interpretation of the sedimentary rocks suggests that they were laid down over hundreds of millions of years, which means that average sedimentation rates throughout this interval must have been very low.

All other things being equal, lower average rates of sedimentation will preserve fewer fossils, so if the standard view is right, the 250,000 fossil species so far documented ought to represent a very low percentage of all the species that have ever existed.

On the other hand, if a significant proportion of the sedimentary rocks and their enclosed fossils were deposited rapidly, for example by the Flood, a much larger percentage of species should have been captured in the fossil record.

Observations suggesting that a very large percentage of modern species have a fossil record (see right) are consistent with the creationist claim that the sedimentary layers were deposited much more rapidly than the conventional interpretation suggests.

Percentage of modern bivalves and gastropods (Point Conception, California, USA) with a fossil record:

Bivalve families	90.6%
Bivalve genera	84.2%
Bivalve species	80.1%
Gastropod families	88.3%
Gastropod genera	81.8%
Gastropod species	75.6%

Percentage of terrestrial vertebrates with a fossil record:

Terrestrial vertebrate orders	97.7%
Terrestrial vertebrate families	79.1%
Terrestrial vertebrate families (excluding birds)	87.8%

PRESERVATION OF SOFT TISSUE IS EASIER TO EXPLAIN BY RECENT BURIAL

In 2005 news headlines proclaimed the discovery of blood vessels and red blood cells, along with connective tissue and bone cells, in a *Tyrannosaurus rex* leg bone. Conventional dating methods estimated this bone to be 68 million years old. Later, intact proteins were also identified.

This find was so surprising that scientists were skeptical. It is difficult to conceive of soft tissues and proteins surviving for tens of millions of years. Some argued that the dinosaur soft tissues were actually organic films produced by bacteria after fossilization.

The scientist who made the original discovery, however, demonstrated that these truly were tissues and organic materials original to the dinosaur. Subsequent studies of an even older hadrosaur fossil (conventionally dated to 80 million years ago) confirmed the presence of soft tissues, including proteins and cell types found only in vertebrates. Most recently, traces of what may be genetic material have been identified in *Hypacrosaurus* bones from Montana, estimated to be about 75 million years old.

The presence of these soft tissues seems easier to explain if the fossils are only thousands of years old.

CAN WE LOCATE THE END OF THE FLOOD IN THE ROCK RECORD?

The most likely location of the end of the Flood in the rock record is at the top of the Mesozoic layers, where Cretaceous rock layers give way to Paleocene rock layers. Several lines of evidence seem to support this conclusion:

1. Sedimentary rock formations belonging to the Paleozoic and Mesozoic portions of the rock record are often very widespread, even spanning continents. This is consistent with sediment deposition during the global Flood. Sedimentary rock formations deposited after the Mesozoic, however, are much smaller in scale (regional or even local), which would be consistent with post-Flood processes.

2. By carefully studying the sedimentary layers, geologists can often work out the direction of the water currents that deposited them. In the Paleozoic and Mesozoic layers, the currents seem to have been remarkably consistent across continents, irrespective of the topography (shape of the landscape). This can be explained if the continents were submerged under water at that time. But in layers formed after the Mesozoic, the water current directions become more scattered, which would again be consistent with post-Flood processes.

3. Since it is unlikely that animals spreading out from the landing site of the ark would have returned to the exact locations where their ancestors lived before the Flood or were buried during the Flood, we would expect to see a major break in the mammal fossil record associated with the end of the Flood. Studies indicate that such a break occurs at the top of the Mesozoic.

THE FLOOD ENDS

Eventually the complete replacement of the old pre-Flood ocean floor brought the Flood to an end. Since the new ocean crust was relatively warm, it had less of a density difference with the underlying mantle and, therefore, less of a tendency to sink back into the earth's interior.

The process of subduction slowed down and eventually came to a virtual stop. This, in turn, brought an end to the overturn of the mantle and the process of rapid seafloor spreading. The fountains of the great deep closed up and the global rain ceased.

As the new ocean crust cooled and subsided, the waters drained off the continents and back into the deepening ocean basins. The Flood finally came to an end.

WHERE DID THE ARK LAND?

The short answer is that we do not know for certain! The Bible is not specific, telling us only that the ark rested upon "the mountains of Ararat" (Genesis 8:4). This indicates a region and not a particular mountain.

The traditional landing site is the peak known as Agri Dagh in eastern Turkey. This mountain, however, is a volcanic cone that seems to have formed some time after the Flood. It sits on top of fossil-bearing sedimentary rocks that were deposited by the Flood.

Other possible landing sites in the region have been explored, but searches have failed to come up with convincing evidence of the ark. It seems unlikely that any traces of a wooden structure like the ark would remain after 4,300 years or more.

4

A NEW WORLD EMERGES

The Flood changed the world in dramatic ways. Many communities of plants and animals did not recover. Others multiplied and diversified to produce the living communities we see today. Over time, the world dried and cooled, leading to the advance of ice sheets across the northern continents. Humans spread out from Babel, transitioning from Old Stone Age cultures to New Stone Age cultures and beyond.

AN UNFAMILIAR WORLD

Upon leaving the ark, Noah and his family were commanded by God to be fruitful and increase in number and fill the earth (Genesis 9:1). The former world, however, had been destroyed, so the descendants of Noah faced much that was unfamiliar (2 Peter 3:6).

The landscape had been utterly transformed—lakes, rivers, mountains, and even the continents and oceans, were different. Furthermore, the upheaval of the Flood had brought about ongoing geological and climatic instability. This was now a world of explosive volcanoes, devastating earthquakes, and whirling hurricanes.

This may help explain the refusal of Noah's descendants to obey God's command to disperse. Perhaps they doubted God's covenant promise that he would never again send "a flood to destroy the earth" (Genesis 9:11). Whatever the reason, while the animals were repopulating the new world, the people decided to settle on the plain of Shinar and set about building a city and a tower (Genesis 11:1-4).

Noah and his family supervise the mammals and birds as they leave the ark. Among the ark animals shown are:

- **A** *Gastornis* (representing the now-extinct gastornithids).
- **B** *Darwinius* (representing the now-extinct adapids).
- **C** *Palaeomastodon* (representing the now-extinct Palaeomastodontidae).
- **D** *Parapropalaeohoplophorus* (representing the armadillos).
- **E** *Mesonyx* (representing the now-extinct mesonychids).
- **F** *Protylopus* (representing the camels).
- **G** *Hyracotherium* (representing the horses).
- **H** *Eotitanops* (representing the now-extinct brontotheres).
- **I** *Proailurus* (representing the cats).
- **J** *Prodinoceras* (representing the now-extinct uintatheres).
- **K** *Protoceras* (representing the now-extinct protoceratids).
- **L** *Icaronycteris* (representing the now-extinct icaronycterids).
- **M** *Phenacodus* (representing the now-extinct phenacodontids).
- **N** *Palaeotragus* (representing the giraffes).
- **O** *Eotragus* (representing the bovids).
- **P** *Palaeotis* (representing the ostriches).
- **Q** *Leptictidium* (representing the now-extinct leptictids).
- **R** *Conuropsis* (representing the parrots).
- **S** *Eostrix* (representing the owls).
- **T** *Piculoides* (representing the woodpeckers).

DEVASTATED BIOMES: CASUALTIES OF THE FLOOD

The Flood destroyed the created biomes that had once dominated large regions of the former world, and many never recovered.

One casualty was the floating forest, which was probably unable to regrow in the stormier waters of the post-Flood world. Most of the plants and animals that inhabited this biome quickly became extinct.

In other cases, a few survivors managed to persist where conditions were ideal. The continent-fringing stromatolite reef did not recover. But stromatolites are still able to grow today in some sheltered spots where the waters are salty enough to inhibit grazing.

Furthermore, creatures that had once lived in separate biomes now came into contact with one another, meaning that there were casualties of competition. For example, the gymnosperms may have been outcompeted by the more rapidly reproducing flowering plants. In turn, this may have been a significant factor in the extinction of the dinosaurs, which had thrived in the gymnosperm-dominated habitats of the pre-Flood world.

MULTIPLYING AND FILLING THE EARTH

Today we see the result of the animals multiplying and spreading across the earth: a great variety of animals filling many different habitats. But how did the animals reach the places where we find them today?

We know that by the end of the Flood the earth's continents were in roughly their present positions. Some landmasses were separated by wide oceans—such as the Americas, separated from Europe by the Atlantic Ocean. Others were surrounded by water—such as Australia and many islands.

Animals must have reached these places by crossing seas and oceans. Hitching a ride on floating mats of vegetation is one way they may have done this.

During the Flood, forests were ripped up and swept into the oceans. Some of the vegetation sank quickly and was buried. But some stayed afloat for centuries. This floating vegetation formed large mats, a bit like those observed when tsunamis or tidal waves hit modern-day coastlines—but much larger.

After the Flood, plant-eating animals would have been attracted onto these mats by the food they provided. Meat-eaters would have followed the plant-eaters. Ocean currents would have moved these mats around, taking the animals with them.

Transport on floating mats helps explain some otherwise puzzling plant and animal distributions. For example, some types of plants and animals are found in limited areas on opposite sides of an ocean, but nowhere else. These distributions are known as **disjunct ranges**. In most cases, major ocean currents connect the two areas. Rafting across the ocean might explain these distributions.

Rafting might also explain the isolation of animals such as marsupials on island continents. Marsupials reproduce more quickly than placental mammals and do not need to invest so much time in the care of their young. A marsupial mother is able to nurture an infant out of the pouch while still suckling, an infant in the pouch, and a baby in the womb—all at the same time. So perhaps the marsupials were able to raft to places like Australia ahead of the more slowly reproducing placentals. By the time the placentals had reached the same points, the plant rafts had already degraded.

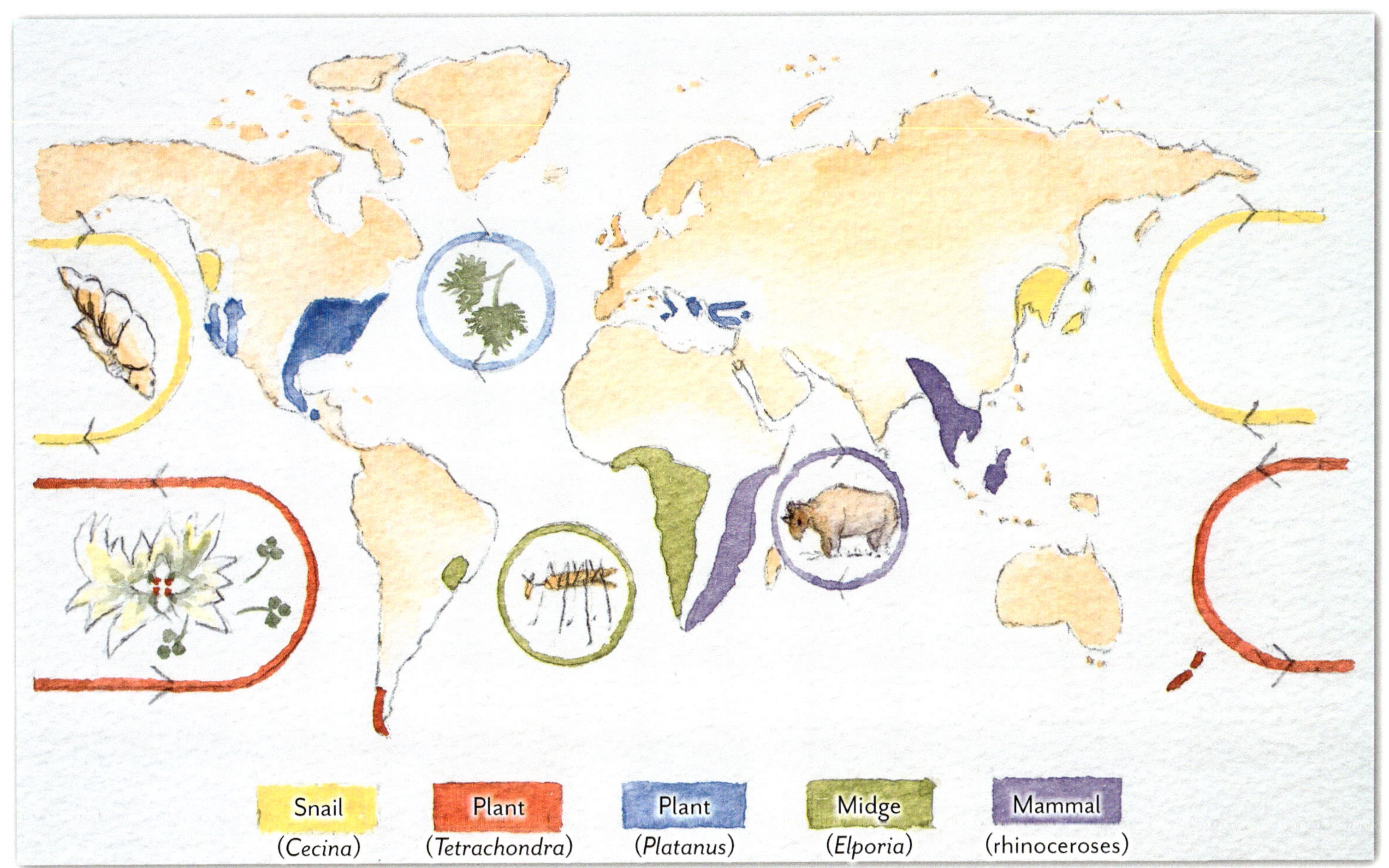

▲ Rafting on vegetation mats that drifted with the ocean currents can explain the disjunct (or split) ranges of many plants and animals. Note that modern distributions are shown. There are also fossils of *Platanus* in western and central Europe, which put it closer to the Atlantic for crossing the ocean in the early post-Flood period.

SURVIVING THE FLOOD OUTSIDE THE ARK

Of course, only the birds and air-breathing land animals needed to spread out from the place where the ark landed. Aquatic animals and most plants repopulated the world from the places where they had survived outside the ark.

Many aquatic animals had perished in the turbulent waters of the Flood, but sufficient numbers survived to refill the devastated oceans, lakes, and rivers. Saltwater and freshwater fishes may have survived the Flood in different parts of the water column. Freshwater is not as dense as saltwater, so it tends to float at the top. In this way, zones of freshwater and saltwater could have developed during the Flood, especially considering the amount of rain that must have fallen. Also, given the changes that have taken place within the kinds since the Flood, some creatures may at that time have had a wider range of salt tolerances than their modern descendants.

Some plants were taken onto the ark for food, but most survived outside the ark. Many plants would have been able to survive as seeds, airborne particles, or as part of the floating vegetation mats. Insects could also have survived as eggs, larvae, or even adults attached to the vegetation.

THE ARK KINDS DIVERSIFY

At the time of the Flood, a single pair (in the case of the "unclean" animals) and seven pairs or seven individuals (in the case of the "clean" animals) represented each of the created kinds of birds and land animals. From these survivors has come the diversity that we see in their descendants today.

Let us consider the implications of this with three examples.

Cats. It seems likely that all members of the cat family—including lions, tigers, leopards, pumas, lynxes, and domestic cats—arose from an original pair of cats that were taken on board Noah's ark. Many modern cat species can produce crosses, strong evidence that they belong to the same created kind. For example, lions can cross with tigers, leopards, and jaguars; leopards can cross with jaguars and pumas; lynxes can cross with domestic cats. The ancestral ark cat may have resembled *Proailurus*, the earliest cat to appear in post-Flood rocks.

DID YOU KNOW?

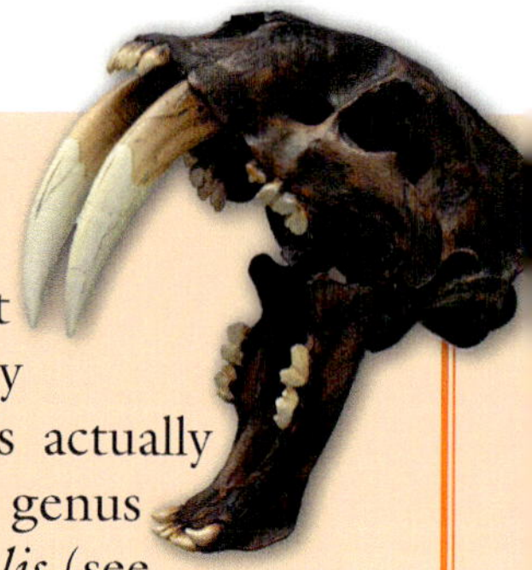

The ancient big cat *Smilodon*, commonly called the saber-toothed tiger, is actually not a tiger. Tigers belong to the genus *Panthera*. But along with *Dinofelis* (see below) *Smilodon* was a member of a separate subfamily of saber-toothed cats, all now extinct.

▼All members of the cat family (Felidae), including modern species such as tigers, leopards, and domestic cats, as well as extinct types such as the saber-toothed *Dinofelis*, were descended from the ancestral cat on the ark. This ancestral cat may have resembled *Proailurus*, a fossilized cat known from Upper Oligocene to Lower Miocene rocks.

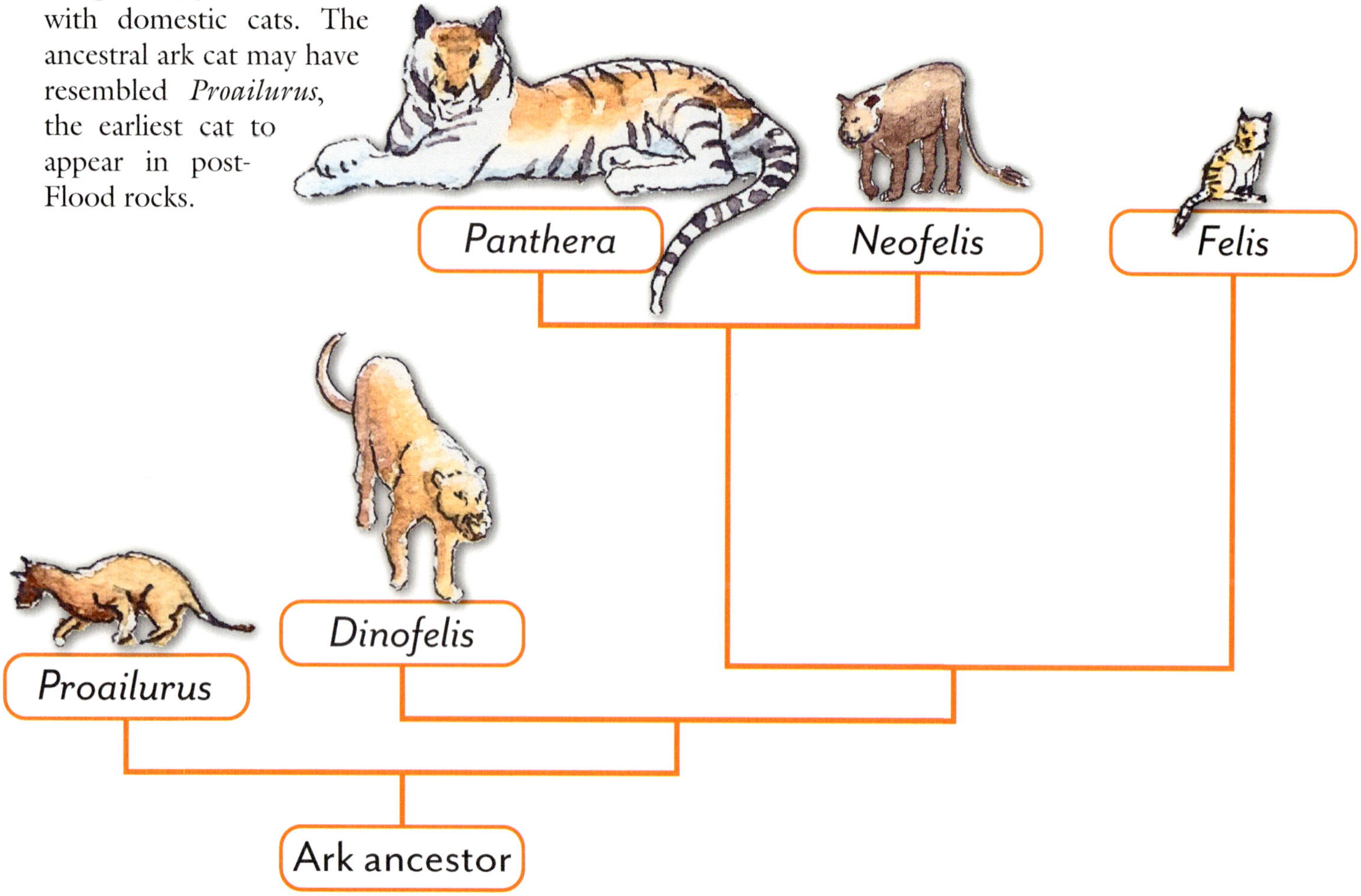

Horses. About 150 species of horses lived after Noah's Flood; we find their fossil remains buried in sediments laid down in post-Flood times. All of these species seem to belong to the same created kind, which means they must have arisen from one pair of horses Noah took on board the ark. The ancestral ark horse probably resembled *Hyracotherium*, the earliest horse to appear in post-Flood rocks.

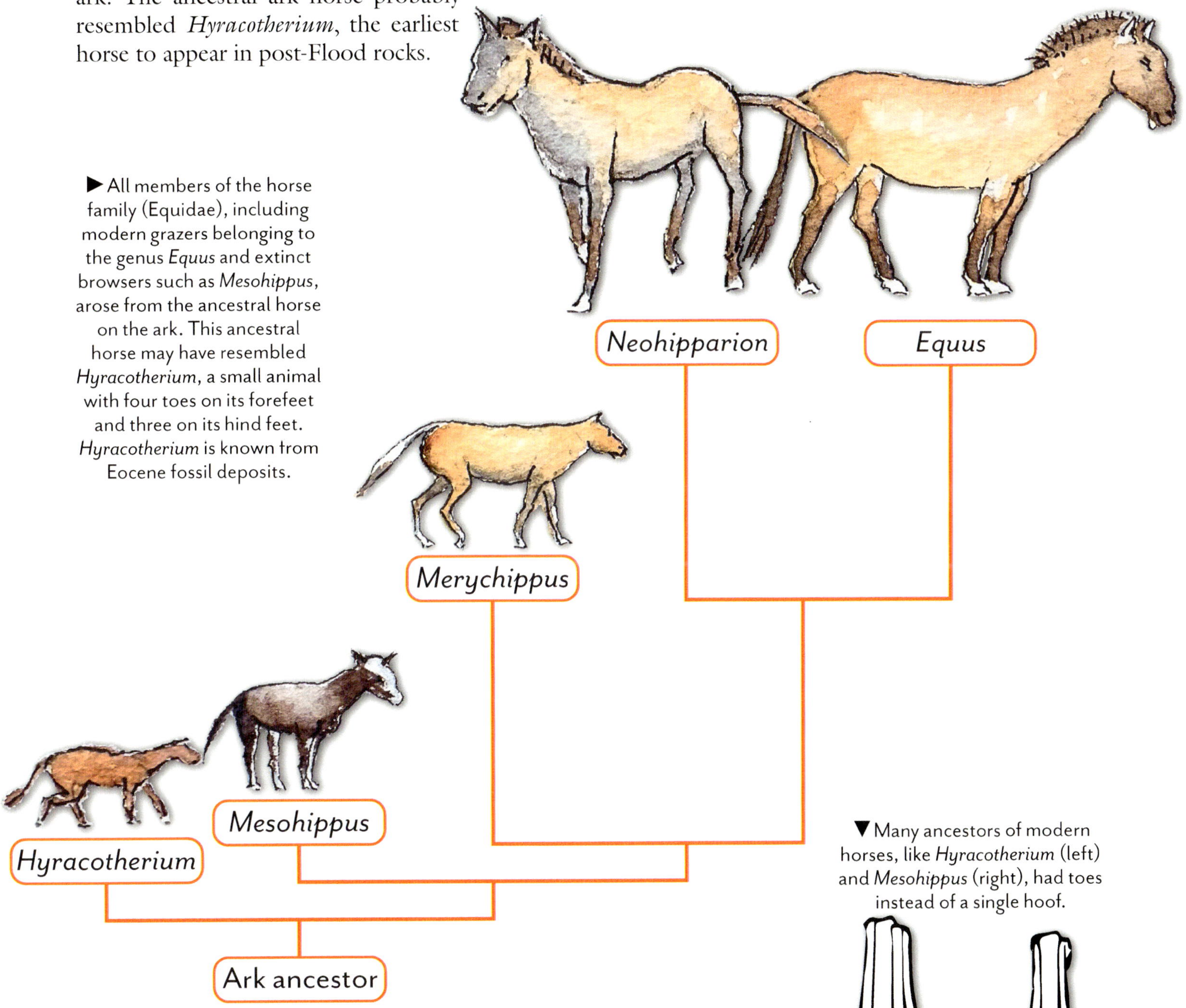

▶ All members of the horse family (Equidae), including modern grazers belonging to the genus *Equus* and extinct browsers such as *Mesohippus*, arose from the ancestral horse on the ark. This ancestral horse may have resembled *Hyracotherium*, a small animal with four toes on its forefeet and three on its hind feet. *Hyracotherium* is known from Eocene fossil deposits.

Neohipparion
Equus
Merychippus
Mesohippus
Hyracotherium
Ark ancestor

▼ Many ancestors of modern horses, like *Hyracotherium* (left) and *Mesohippus* (right), had toes instead of a single hoof.

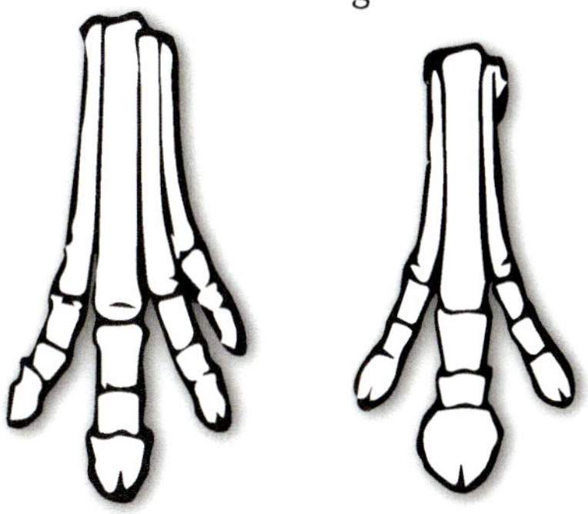

HOW MUCH DIVERSIFICATION HAS OCCURRED SINCE THE FLOOD?

One way to estimate how much **diversification** has taken place since the Flood is to consider the number of different species within each family of air-breathing land mammals. It turns out that most of these mammal families contain few species, and only a few of them are large with many species.

Making the reasonable assumption that the family is roughly equivalent to the created kind, it seems that most kinds have undergone only modest diversification since the Flood. But some kinds (including many families of bats and rodents) have hundreds of species and have clearly diversified to a much greater extent.

Brontotheres. This is an extinct group of rhino-like mammals with elongated skulls, characteristic upper molars, and (in the larger forms) paired horns on the snout. Some were true giants, such as *Megacerops*, which was about 8.2 feet (2.5 meters) tall at the shoulders and probably weighed over 3.3 tons (3 tonnes). These giants seem to have arisen from a much smaller ancestral ark brontothere resembling *Eotitanops*, the earliest brontothere to appear in post-Flood rocks.

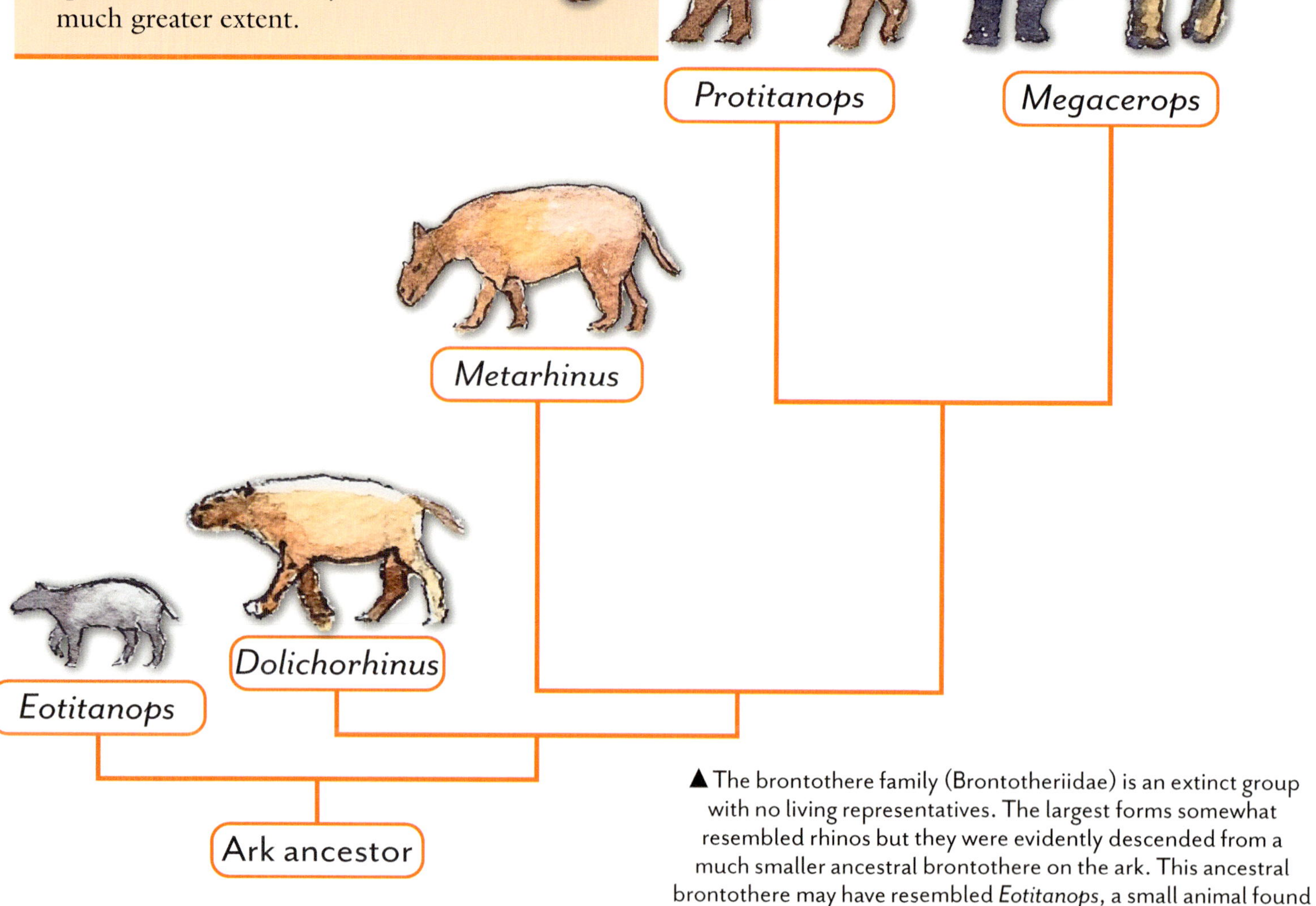

▲The brontothere family (Brontotheriidae) is an extinct group with no living representatives. The largest forms somewhat resembled rhinos but they were evidently descended from a much smaller ancestral brontothere on the ark. This ancestral brontothere may have resembled *Eotitanops*, a small animal found in Eocene fossil beds that stood only 1.5 feet (45 centimeters) at the shoulder.

So it seems that God equipped the created kinds with the potential for significant change. He designed them with the ability to bring forth many new species and varieties. This process has been called diversification. This in-built ability to adapt can be seen as the Lord's provision to enable the created kinds to survive in a changing world—especially after the global restructuring that took place at the time of Noah's Flood.

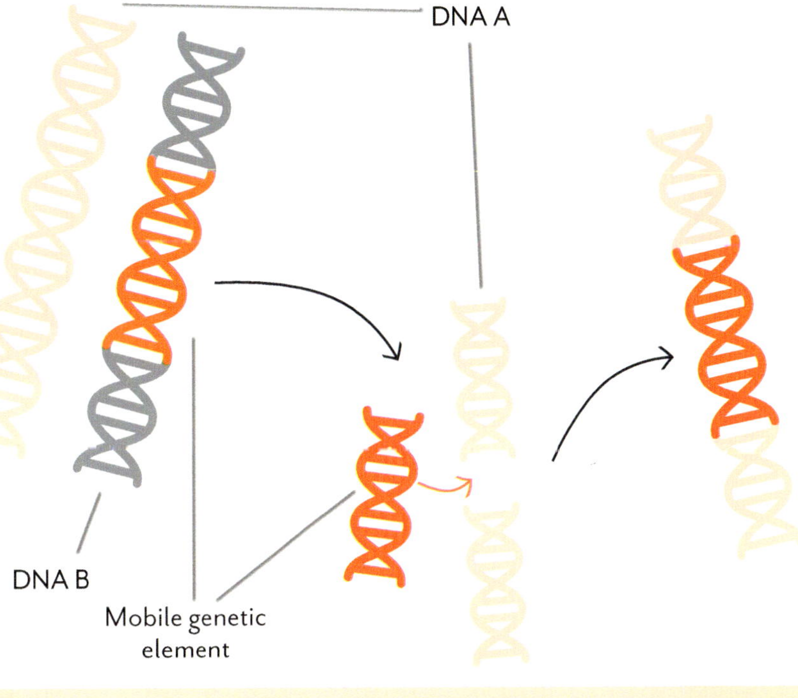

▲A mobile genetic element jumping from one piece of DNA to another. Processes like this may have contributed to the rapid diversification of animals and plants after the Flood.

HOW DID SO MUCH CHANGE HAPPEN SO QUICKLY?

We know from the Bible that many modern species were around within 1,000 years or less from the time of Noah's Flood, including lions (Job 4:10-11; 10:16; 28:8; 38:39), camels (Genesis 12:16), and donkeys (Genesis 12:16). This means that the birds and land animals must have diversified very quickly after their ancestors stepped off the ark.

This raises an obvious question: *How could so much change have happened so quickly?* Part of the explanation may involve the activity of **mobile genetic elements**. It has been known for several decades that some pieces of the genome can independently replicate themselves and move around. Today these mobile genetic elements are mostly regarded as harmful parasites, but perhaps God originally designed them to co-operate with the genes of living organisms to produce diversity.

There are a number of ways in which they could have done this. For example, we know that some of these mobile elements are able to boost or suppress the activity of genes—even switching on inactive genes and switching off active ones. They can also cause genes to recombine into new arrangements. Furthermore, because these elements are mobile, they can take complete genes and move them around within an organism—or even between organisms! Altering the genetic make-up of organisms in this way can produce substantial change in a very short amount of time.

Harmful mutations—random genetic copying mistakes—occur in these mobile elements when they replicate themselves, causing damage which accumulates over time. This damage may have led to the beneficial function of the mobile elements being lost or reduced. As a result of this degeneration, most of these mobile elements are currently inactive and some are actually harmful.

Other factors are certainly involved as well, and this is an area where more research is needed.

COOLING AND DRYING OF THE WORLD

During the catastrophic rearrangement of the earth's plates during the Flood, an enormous amount of heat must have been generated. Much of this heat was inevitably transferred to the ocean waters, so that by the end of the Flood the oceans were much warmer than they had been before the Flood.

The warmth of the oceans after the Flood led to the evaporation of a great deal of seawater into the atmosphere. As the moist air circulated over the continents, it cooled and condensed as heavy and prolonged rain.

The heavy rainfall after the Flood resulted in erosion and sedimentation on a massive scale. This helps account for many features of the earth's surface, including widespread erosion surfaces called peneplains, thick wedges of river-deposited material called alluvial fans, and the extensive delta deposits found near the mouths of many modern rivers.

In addition, large lakes, some of which no longer exist, developed in many areas. The famous Fossil Lake of the Green River Basin in Wyoming, USA, is one example. This lake seems to have formed within only a few decades of the end of the Flood.

The climate in the region around this lake at that time was temperate to subtropical, so the

vegetation was lush with many warmth-loving plants. Many types of fishes (such as *Diplomystus* and *Knightia*) and reptiles (such as the turtle *Echmatemys*) swam in the lake. Birds (including *Presbyornis*) and mammals (such as *Uintatherium*) thronged the shores. Invertebrates such as snails and insects were abundant and mound-shaped stromatolites grew around the lake margins.

As the oceans cooled, the rate of evaporation declined and the earth began to dry out. Grasslands expanded across the continents and deserts began to form. In the mid to high latitudes the forests changed from subtropical to warm temperate to cool temperate and eventually to tundra. The stage was set for the **ice age**.

THE GREEN RIVER FORMATION: FORMED IN POST-FLOOD LAKES

Much evidence indicates that the Green River sediments were deposited in post-Flood lakes:

1. The formation sits on top of widespread marine sediments that date from Noah's Flood and is separated from them by an erosion surface that was probably carved by the draining floodwaters.
2. The Green River sediments are enclosed within natural hollows (basins), just like the sediments building up in modern lakes.
3. The Green River sediments are distributed in a bullseye pattern, with coarser material at the edges of the basin and finer material in the center, just as in modern lakes.
4. The Green River sediments contain the fossilized remains of freshwater animals and plants—consistent with a lake ecosystem.
5. Bird tracks and nests, caddis fly mounds, and stromatolites occur around the edges of the formation, as expected of lake shorelines.
6. The preservation of the fossilized fishes indicates they were living in shallow water near the edges of the basin and in deeper water towards the center.

HUMANS SPREAD ACROSS THE EARTH

For some time, Noah's descendants refused to obey God's command to fill the earth. Instead, they devoted themselves to building the tower of Babel at Shinar. But four generations after the Flood, in the days of Peleg, "the earth" was divided (Genesis 10:25)—in context referring to the world of people rather than the physical globe. Genesis 11:1-9 tells us that God came down to confuse their language and cause them to scatter as he had always intended.

Small groups, perhaps tribes or families, set out from Babel and began to disperse into Africa, Asia, and Europe. But life was not easy for these migrating peoples. They were forced to adopt a primitive lifestyle. Before they were able to find bodies of ore and extract metals from them, they turned to stone tools. Before they were able to settle down and build cities they sought shelter in caves and tents. Before they were able to manipulate their environment and become farmers they survived by hunting and gathering.

In other words, these pioneers lived as we know Paleolithic (Old Stone Age) people did. But over time there was a general transition to more advanced cultures. This transition explains the progression in the archaeological record from Paleolithic (Old Stone Age) cultures to Neolithic (New Stone Age) cultures and beyond.

A particularly remarkable thing about these early post-Babel populations is how variable

they were, much more so than modern humans. One of the oldest of these populations lived at the site now known as Dmanisi in the former Soviet republic of Georgia—less than 200 miles (320 kilometers) from the traditional site of the mountains of Ararat. This location is one of the richest sites of human fossils ever discovered. But the human fossils have proved difficult to classify because they display a strange mix of primitive and advanced features.

The Dmanisi humans had small brains (550 to 775 cubic centimeters) and somewhat ape-like shoulders and arms, but their legs and feet were more modern in appearance. They are found in association with stone tools—mostly choppers and scrapers—of a type known as Oldowan, after the Olduvai Gorge in Tanzania where they were first described. Some scientists think these fossils represent a new species (*Homo georgicus*), while others suggest that they are a type of *Homo erectus*.

At the time these people were living at Dmanisi, the surrounding area was a mix of woodland and open grassland. Large carnivores stalked the landscape, including saber-toothed cats (*Megantereon*) and wolves (*Canis*). There were scavengers too, including hyenas (*Pachycrocuta*). Large herbivores were also present, including elephants (*Archidiskodon*) and rhinoceroses (*Dicerorhinus*). Medium-sized herbivores included horses (*Equus*), deer (*Cervus*), and goat antelopes (*Soergelia*), and there was a host of smaller mammals. Giant ostriches (*Struthio*) also roamed the grasslands, some individuals reaching up to 12 feet (3.7 meters) tall.

EARLY POST-BABEL HUMAN DIVERSITY

The range of sizes and shapes of skulls and bodies found among early post-Babel humans is remarkable.

In Asia and Africa there were tall, slender humans with rather small brain sizes—*Homo erectus* (about 1,100 cubic centimeters) and *Homo ergaster* (about 900 cubic centimeters). In Europe there were short, stocky humans with large brain sizes—*Homo neanderthalensis* (about 1,600 cubic centimeters). And on the Indonesian island of Flores there were tiny humans with tiny brain sizes—*Homo floresiensis* (about 400 cubic centimeters).

Most recently, 15 skeletons found in the Rising Star cave near Johannesburg in South Africa have added *Homo naledi* to the human family (with a brain size of about 500 cubic centimeters).

These brain sizes compare with the modern human average of about 1,200 cubic centimeters.

The skulls found at Dmanisi are especially noteworthy since they are so different from one another, and yet they come from a single locality and date to about the same time. It seems that early post-Babel human populations were much more diverse than modern human populations.

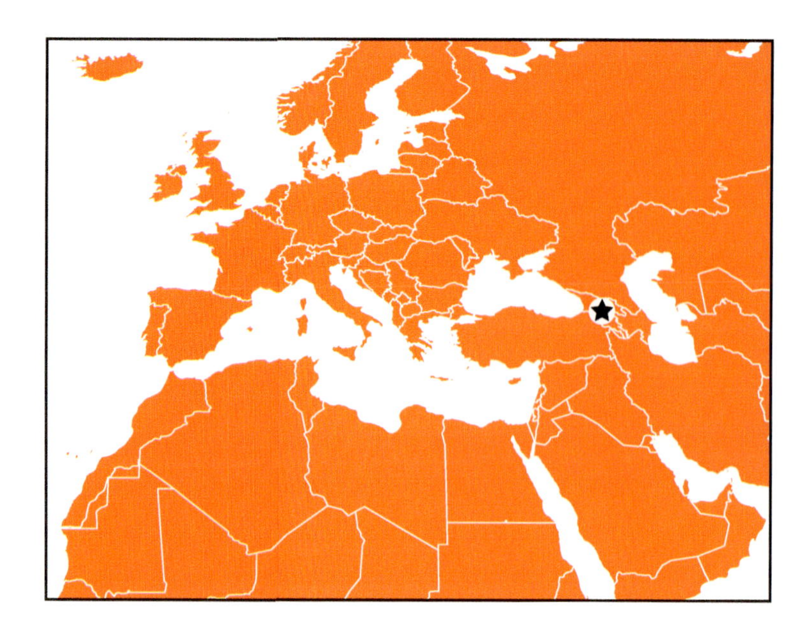

▲ Site of the Dmanisi archaeological dig in eastern Europe.

▼ Five skulls from Dmanisi, showing the extreme variation in a single early post-Babel population:

- A Skull D2280.
- B Skull D2282.
- C Skull D2700.
- D Skull D3444.
- E Skull D4500.

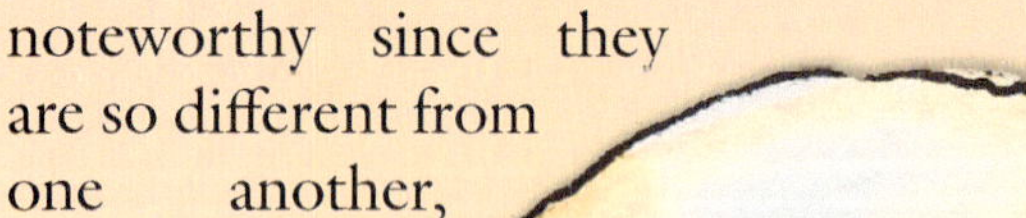

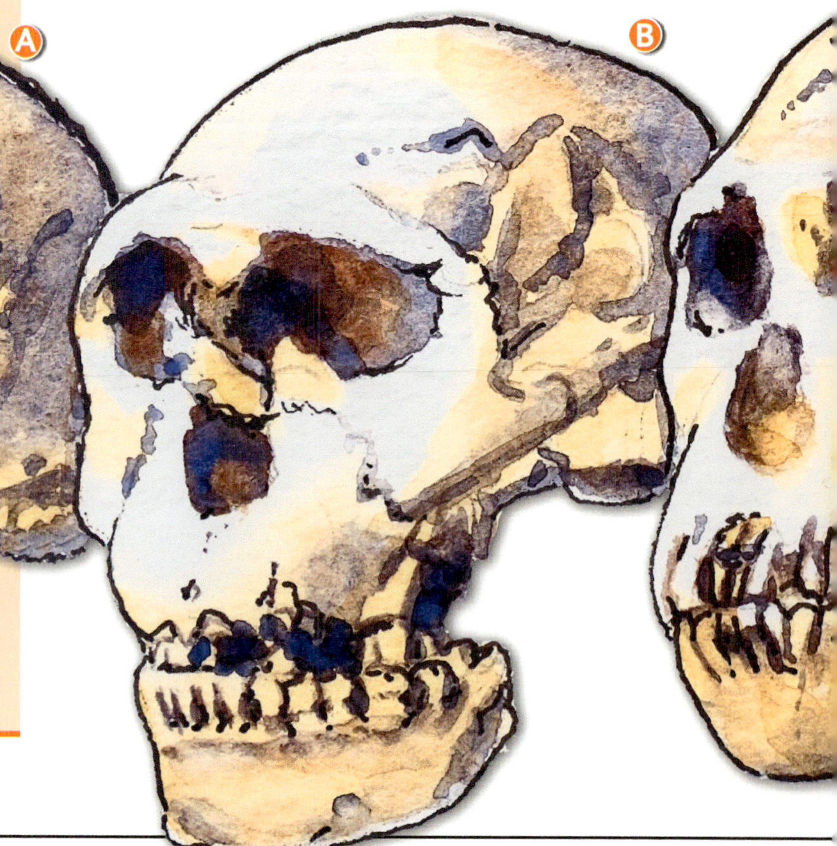

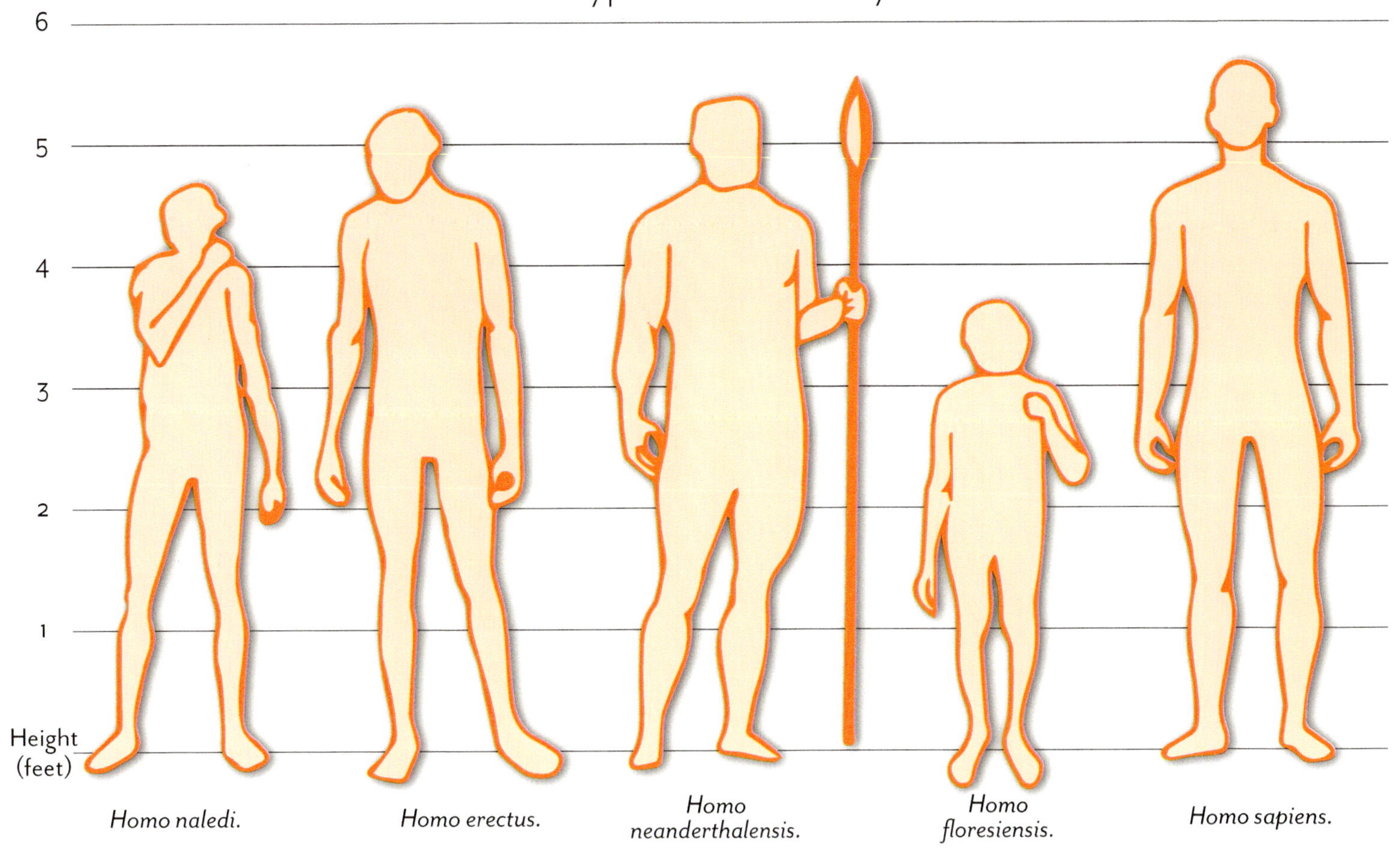

Early post-Babel human diversity.

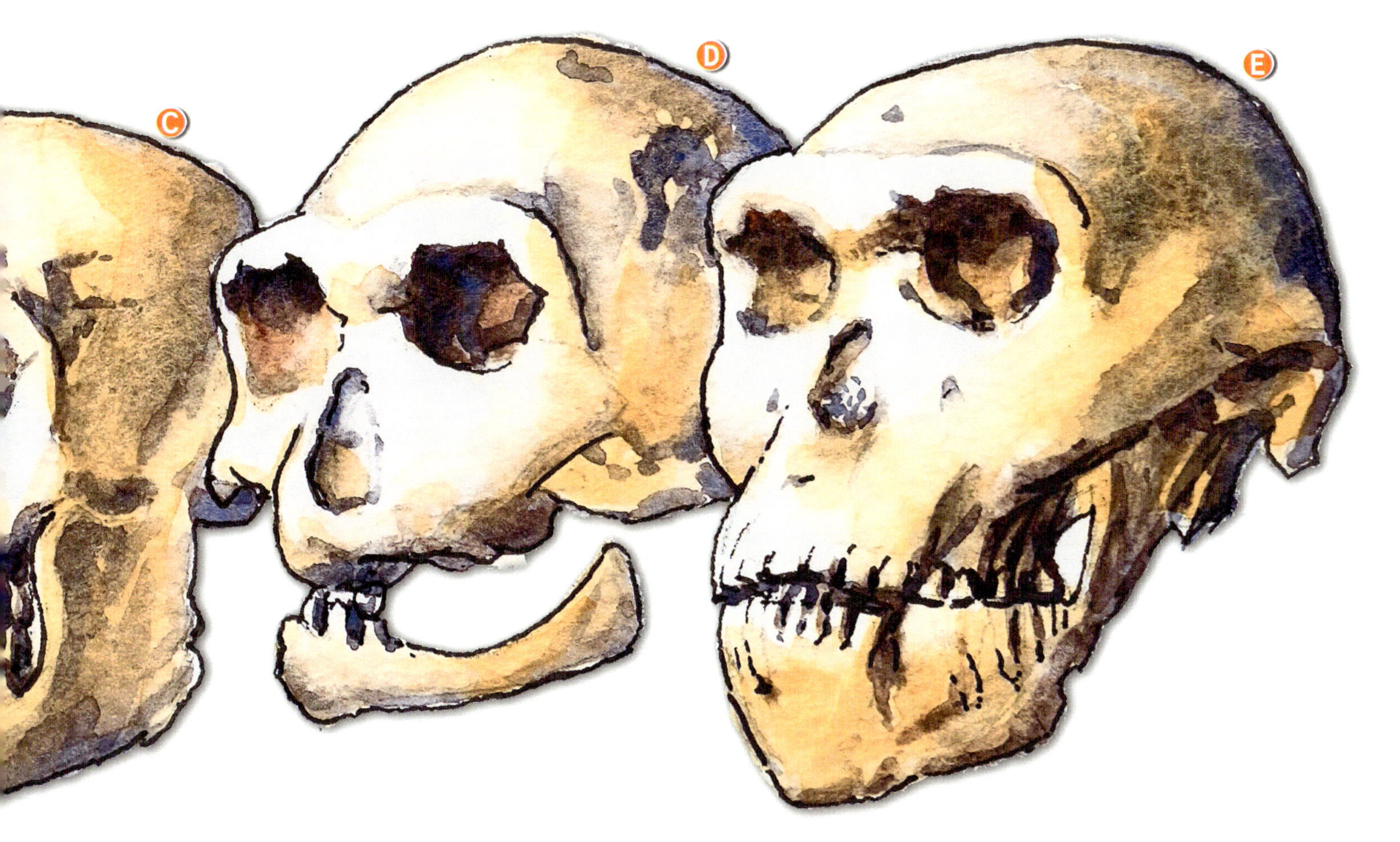

THE ICE ADVANCE

By the time the people were spreading from Babel, great ice sheets had started to develop across much of the northern hemisphere. Perhaps this was part of God's plan to prevent people from quickly resettling in one place again.

As a consequence of the ice buildup, the global sea level dropped and areas of former seafloor were exposed. The Baltic and North Seas became dry land, and Asia and North America were connected across what had once been the Bering Strait. These land bridges provided passageways

HOW DID THE ICE ADVANCE HAPPEN?

As the warm post-Flood oceans cooled by evaporation, the moisture in the atmosphere eventually fell in the mid and high latitudes as snow. Storm after storm developed along the coastlines of Europe, Asia, and North America. The result was a snowblitz—with ice sheets building up across large areas within only centuries.

The advance of the ice was halted when the temperature of the ocean reached a level at which it could no longer sustain the global buildup of ice. The amount of evaporation from the ocean surface declined, and the ice sheets stopped growing and started melting. From start to finish, the ice advance may have lasted only a few centuries.

for the migration of people and animals into new areas.

Life around the southerly margins of the ice sheets was tough. In Britain, the ice sheets extended as far south as London and the Thames Valley. Beyond the **glacial** ice, there was a tundra-like landscape of grasses, lichens, low bushes, and stunted trees. Large patches of ground remained permanently frozen all year round.

Groups of Neanderthal hunters tracked large herbivores across this landscape. The Neanderthal people were well suited to cold conditions. They were short, stocky, and muscular, with thick, strong bones. Their skulls were low-domed with prominent brow ridges. These features allowed them to conserve heat effectively, and their broad noses helped warm the icy air before it reached their lungs.

Large mammals populating this region included mammoths (*Mammuthus*), giant Irish elk (*Megaloceras*), and woolly rhinoceros (*Elasmotherium*). Horses (*Equus*) and reindeer (*Rangifer*) roamed the grassy plains while caves provided dens for bears (*Ursus spelaeus*). Carnivores such as wolves (*Canis*) and lynx (*Lynx*) stalked their prey, including small mammals such as the snow hare (*Lepus*) and birds such as the raven (*Corvus*).

WHY WERE THE ICE AGE MAMMALS SO BIG?

Many ice age mammals were very large—an obvious example being the woolly mammoth (*Mammuthus primigenius*), which stood 10 feet (3 meters) tall at the shoulder and weighed over 6.5 tons (6 tonnes). Its large size can be seen as a design for cold climates, giving it a smaller relative surface area (compared to volume) through which to lose heat. Other traits suitable for life in cold conditions included thick fur, fatty layers under its skin, and smaller ears than modern elephants.

As the ice sheets melted, the woolly mammoth and most of the other large ice age mammals became extinct. The changing climate may have been a source of environmental stress, perhaps helped along by man's hunting activities. Among the last holdouts were the dwarf mammoths of Wrangel Island in the Arctic Ocean. By the time these animals disappeared, the pharaohs were ruling in Egypt.

▲ Columbian mammoths (*Mammuthus columbi*) like this one inhabited the open grasslands of North America as far south as Costa Rica during the ice age. However, the woolly mammoth (*Mammuthus primigenius*) preferred the Canadian tundra to the north.

THE RISE OF CIVILIZATIONS

After the ice had receded, new civilizations began to be established. The oldest of these was Mesopotamia, followed closely by others in Egypt, then India, China, and South America.

New technologies arose quickly, resulting in architectural marvels such as the Egyptian pyramids and the Babylonian ziggurats. Within a few generations, sophisticated manufacturing and trade networks had been re-established.

In the region of Mesopotamia, several ancient cities were founded. Each was a small fortified settlement ruled by a king, and they were often in conflict with one another.

One of the most magnificent of these city-kingdoms was Ur, now modern-day Tell el-Muqayyar in southern Iraq. Before the Euphrates river changed course, Ur was a prosperous port—a center of shipping and commerce. Archaeological excavations have uncovered many buildings, temples, and tombs, along with thousands of inscribed stone tablets.

Ur was also a center of idolatrous worship, dedicated to the moon god Nanna. Unsavory and licentious practices were commonplace. And yet it was from this pagan city that God called Abram to be the father of the Jewish nation (Genesis 12:1-4).

In obedience to God's command, Abram left Ur with his wife Sarai, father Terah, brothers Nahor and Haran, and nephew Lot to go to a place that God promised to show him (Genesis 11:31).

God later changed Abram's name to Abraham, meaning *father of a multitude* (Genesis 17:5). God's plan was to raise up through Abraham a people for himself: the nation of Israel. God promised Abraham that he would make him "a great nation," and that through him "all families of the earth [would] be blessed" (Genesis 12:3).

That great promise was ultimately fulfilled in the coming of Abraham's descendant, Jesus Christ, the incarnate Son of God, whose life, death, and resurrection continues to bring life, healing, and hope to all the peoples of the world.

Abraham's trust in God resulted in his justification before God—his faith was counted to him as righteousness (Romans 4:3). For those who have likewise placed their trust in the God of the Bible, Abraham remains an outstanding example of living, saving faith (Galatians 3:7-8).

5

FOSSILS AND THE FOSSIL RECORD

The reconstruction of earth history presented in this book owes much to the insights we have gleaned from the fossil record. Fossils are the remains or traces of creatures that lived in the past, preserved most often in sedimentary rocks. Fossilization is rare, but more likely to occur when creatures are rapidly buried, as would have been the case in the worldwide Flood. The order in which the fossils are found in the stack of rock layers deposited during the Flood reflects the order in which the original created biomes were destroyed and buried, not the order in which those biomes evolved over long ages.

WHAT ARE FOSSILS?

Fossils are among the most crucial pieces of evidence that we can use to reconstruct the story of the early earth. But what are fossils and how were they formed? What types of fossils do we encounter in the rocks?

Fossils are the remains or traces of creatures that lived in the past, in most cases preserved when they were buried in sediments (such as mud or sand) laid down by water.

Most fossils represent the hard parts of once-living things, such as shell or bone. We call these **body fossils**.

Not all fossils, however, represent the actual remains of once-living things. Some are traces of their activity, such as footprints, burrows, eggs, and droppings. We call these **trace fossils**.

▶This A ammonite shell and B fish skeleton are examples of body fossils.

▼The trace fossils shown here include C nests of dinosaur eggs, D the trackway of a theropod dinosaur, E coprolite (fossilized droppings), and F some animal burrows.

The scientific study of fossils is called paleontology and the scientists who study them are called paleontologists. The word *paleontology* comes from the Greek words *palaios* (meaning *ancient*), *ontos* (meaning *being* or *creature*), and *logos* (meaning *study*); so paleontology is the study of ancient life.

Fossils provide scientists with valuable information about creatures that lived in the past, their habits, and their ways of life. They also help us understand environments that vanished long ago.

DID YOU KNOW?

People have not always recognized fossils as the remains of once-living creatures. In medieval times it was widely believed that fossils had been produced by natural forces in the earth. Others thought they had grown from seeds carried down in the rain. But by the eighteenth century most people had come to think that fossils were the petrified remains of animals and plants.

◀This paleontologist is using a geological hammer to excavate some fossilized bones.

FOSSILS AS EVIDENCE OF THE FLOOD

Many leading naturalists in the seventeenth century considered fossils to be evidence of Noah's Flood. One was John Woodward (1665-1728), who thought that the fossils enclosed in the rock layers were animals and plants that had been overwhelmed by the floodwaters.

In the Sedgwick Museum of Geology in Cambridge, England, you can see a reconstruction of John Woodward's study, along with the original collection cases that contained nearly 10,000 rock, fossil, and mineral specimens.

▶John Woodward with some of the fossils from his extraordinary collection.

WHAT CONDITIONS ARE NEEDED FOR FOSSILS TO FORM?

Fossilization is a rare process, but several things make it more likely that a creature will be preserved as a fossil.

Creatures with hard parts (such as shellfish) are more likely to be fossilized than those with only soft parts (such as jellyfish). Soft parts usually rot quickly after a creature dies, although they can be preserved in exceptional cases.

Rapid burial is very important because this prevents scavengers from eating the remains. So fossils are more likely to form in places where sediments are building up (such as in the ocean) than in places where sediments are being worn away (such as on the land).

Burial in a place with low oxygen levels (such as the bottom of a deep lake) helps slow down rotting and makes fossilization more probable.

Finally, good-quality fossils are more likely to be formed when a creature is buried in fine sediment (such as mud) than when it is buried in coarse sediment (such as sand). Fine sediments are better at preserving details of the dead creature.

The worldwide Flood described in the Bible provided ideal conditions for fossils to form. The floodwaters were laden with sediments that trapped and buried millions of creatures when those sediments were deposited.

STAGES IN FORMING A FOSSIL

1. The creature (in this case a squid-like animal called a nautiloid) dies and is buried before its remains can be destroyed.

2. More layers of sediment build up on top of the buried creature.

3. Pressure and minerals carried in groundwater combine to turn the sediments into rock and the creature's shell into a mineralized fossil.

4. The fossilized shell stays in the rock until it is uncovered by erosion or excavation.

HOW ARE FOSSILS PRESERVED?

Fossils can be preserved in many different ways.

For example, a shell may be dissolved, leaving behind an impression of its inside or outside (a **mold**).

Sometimes the void left by the shell is filled with another mineral that takes the shape of the original shell (a **cast**).

In other cases the original material of the shell is gradually replaced with another mineral (**replacement**).

Sometimes empty spaces or pores in wood or bone are filled by another mineral (**petrification**).

Many fossilized plants are preserved when oxygen and water are squeezed out during burial, leaving behind a thin carbon impression (**carbonization**).

Some creatures are even preserved when they become encased in tree sap, tar, or ice (**entombment**).

COLLECTING FOSSILS

Making your own fossil collection can be great fun and may develop into an absorbing hobby. Serious collectors will have a range of tools including hammers, chisels, brushes, and sieves. But you can begin by simply looking out for fossils the next time you visit the beach! Keep notes on what you find and where you find it—and perhaps take some photographs of the location. Wrap your specimens in tissue or newspaper to protect them. You can clean up and identify your fossils when you get them home. Store your collection in cardboard or plastic trays with labels bearing the name of each fossil, the place where you found it, and the rock formation it came from. There are lots of useful maps and guidebooks to help you.

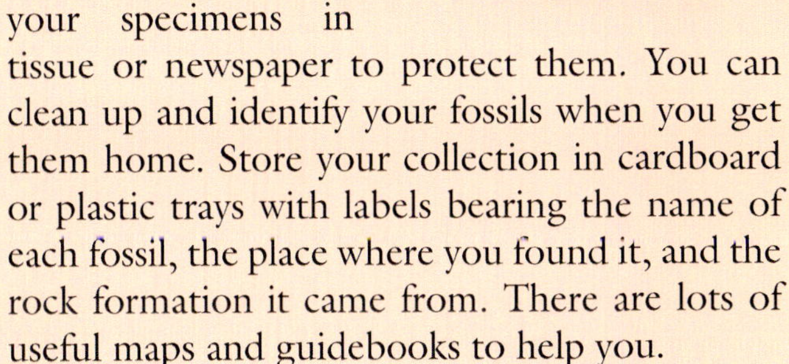

▶ Mold. The outside of a bivalve shell left this concave impression in the sediment.

▲ Replacement. This ammonite shell was replaced with the mineral iron pyrite (often called fool's gold), giving it a shiny appearance.

▲ Carbonization. These fern leaves were preserved as a thin carbon film in dark mudrocks.

▶ Cast. Sediment filled the empty shell of this gastropod so that when the shell was dissolved its coiled shape was preserved.

▲ Petrification. The mineral silica preserved this tree stump by filling in the tiny pore spaces in the wood.

Entombment.
▶ Some beetles captured in tar.
▼ A small flying insect preserved in amber (tree sap)

WHAT KINDS OF ROCKS CONTAIN FOSSILS?

The earth's crust is made up of rocks of various kinds:

- **Igneous rocks**, which are formed when hot molten magmas cool below ground or are erupted as lavas above ground.
- **Sedimentary rocks**, which are made of grains of mud, sand, or gravel carried by water, wind, or ice and deposited in layers.
- **Metamorphic rocks**, which are rocks that have changed their form when subjected to high temperatures or pressures.

Fossils are rarely found in igneous and metamorphic rocks because high temperatures and pressures tend to destroy them. But fossils are common in sedimentary rocks.

Sedimentary rocks build up in layers called **strata**. In an undisturbed pile of sedimentary rocks the first layers to be deposited are at the bottom and the last to be deposited are at the top.

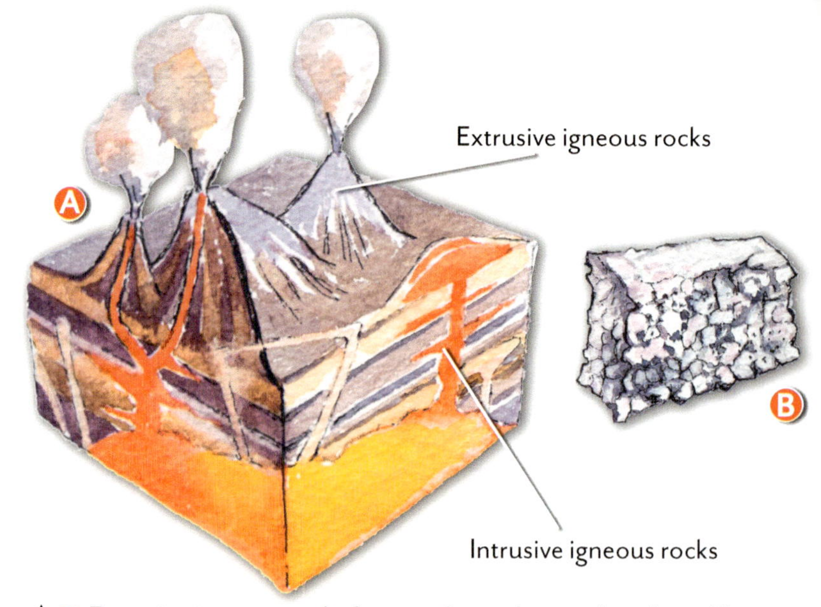

▲ A Extrusive igneous rocks form at the surface and cool quickly to form fine crystals. Intrusive igneous rocks form below the surface and cool more slowly to form larger crystals. B **Granite** is an intrusive igneous rock made of large interlocking crystals of the minerals quartz, feldspar, and mica.

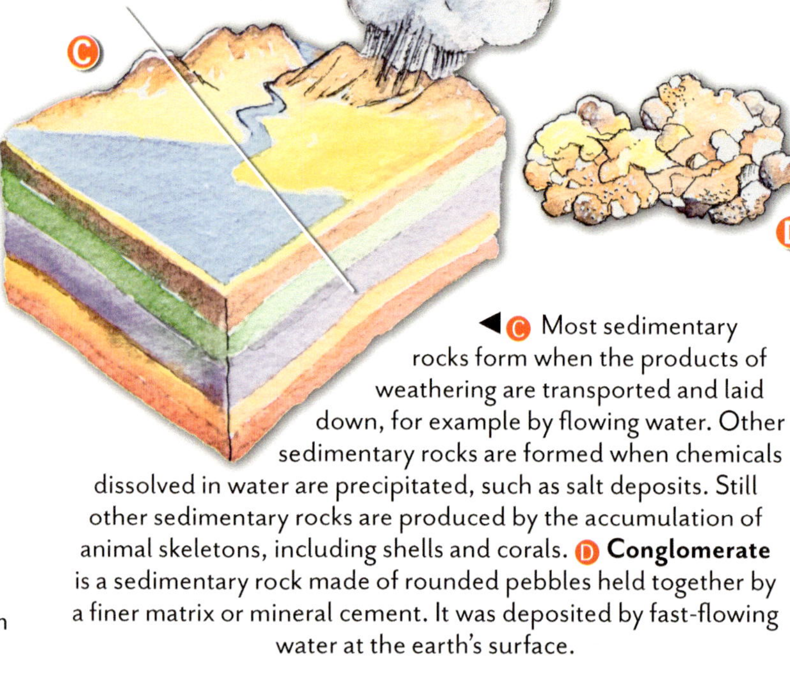

◀ C Most sedimentary rocks form when the products of weathering are transported and laid down, for example by flowing water. Other sedimentary rocks are formed when chemicals dissolved in water are precipitated, such as salt deposits. Still other sedimentary rocks are produced by the accumulation of animal skeletons, including shells and corals. D **Conglomerate** is a sedimentary rock made of rounded pebbles held together by a finer matrix or mineral cement. It was deposited by fast-flowing water at the earth's surface.

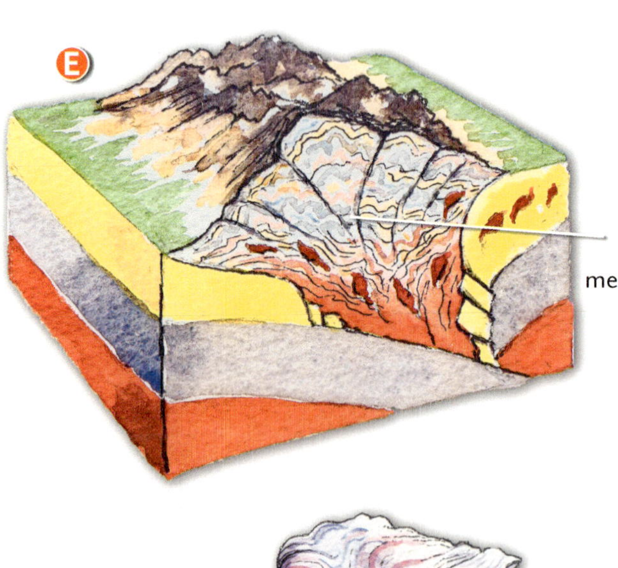

◀ E Metamorphic rocks form when existing rocks are altered by heat or pressure or both. Some rocks are altered locally by the heat of a nearby igneous intrusion, in a process known as **contact metamorphism**. Other rocks are altered over wide areas, for example by deep burial during mountain-building. This process is called **regional metamorphism**. In regional metamorphism the effects of pressure and temperature are usually regarded as of equal weight, but in creationist thinking regional metamorphism is mostly due to elevated temperature and only secondarily to elevated pressure. F **Gneiss** is a metamorphic rock with distinctive light and dark bands. It was formed from a precursor sedimentary rock during regional metamorphism.

Early students of geology recognized patterns in the earth's rock layers. These patterns are summed up in what came to be known as the geological column.

Certain rock types are more common in certain parts of the column. For example, Carboniferous rocks are well known for their coal layers, Permian rocks for their red sandstones, and Cretaceous rocks for their chalk deposits. These rock layers are found in a consistent order the world over.

Fossils buried in these rock layers are also found in a consistent order. Trilobites, for instance, are found only in Cambrian to Permian layers while dinosaurs are found only in Upper Triassic to Cretaceous layers. Trilobites and dinosaurs have not been found together in the same rocks.

There is no place on earth where there is a complete sequence of rock layers. But by studying rocks in many places and matching them up, it is possible to piece together the order in which they formed.

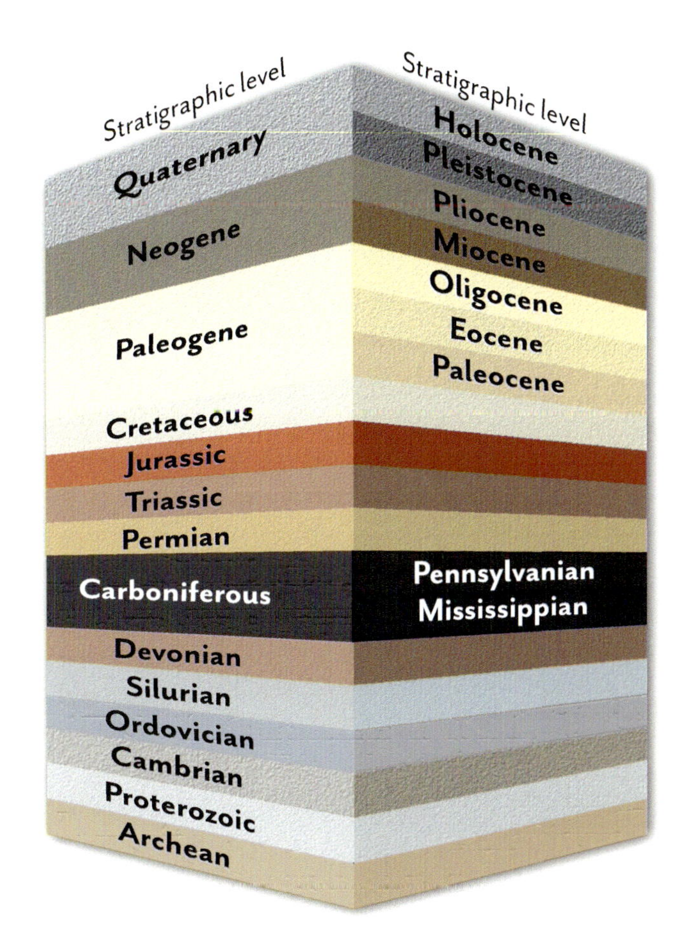

▲The standard geological column has many divisions and subdivisions. Each is characterized by a distinctive assemblage of fossilized organisms.

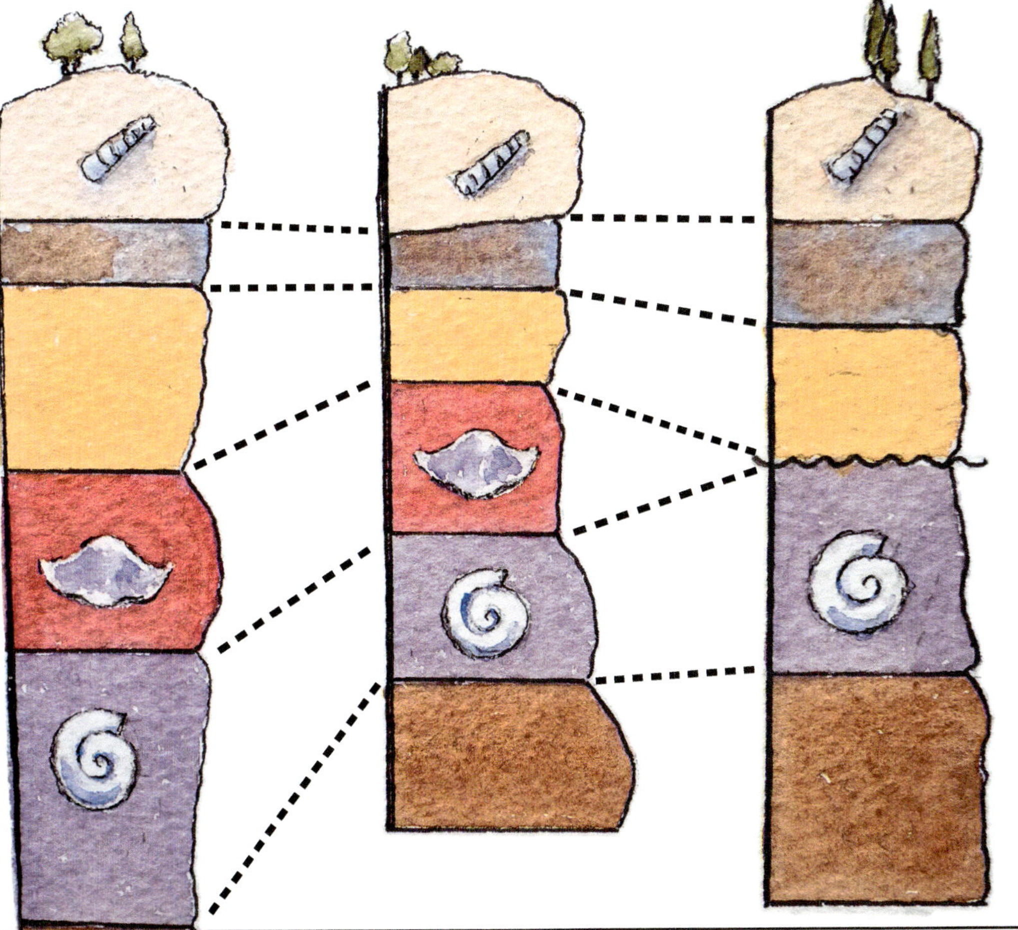

◀Correlation between local rock sequences is achieved by matching up rock types and fossil assemblages.

WHAT DOES THE FOSSIL ORDER MEAN?

Most biblical creationists think that the majority of fossils were formed during the worldwide Flood in the days of Noah. Such a worldwide Flood would have left behind layers of sediment containing the buried remains of many animals, plants, and other creatures.

According to this **creationist** view, the order in which most fossils are found reflects the order in which communities of creatures were buried as the floodwaters rose and covered the earth.

This is in contrast to the **evolutionary** view adopted by most scientists. According to the theory of evolution, gradual changes over many generations led one type of creature to give rise to another. In this way, fishes are said to have given rise to land animals, and reptiles to mammals and birds.

▼ Two interpretations of the geological column. The abbreviation *my* stands for millions of years ago. Modified from Wise and Richardson (2004, p.132).

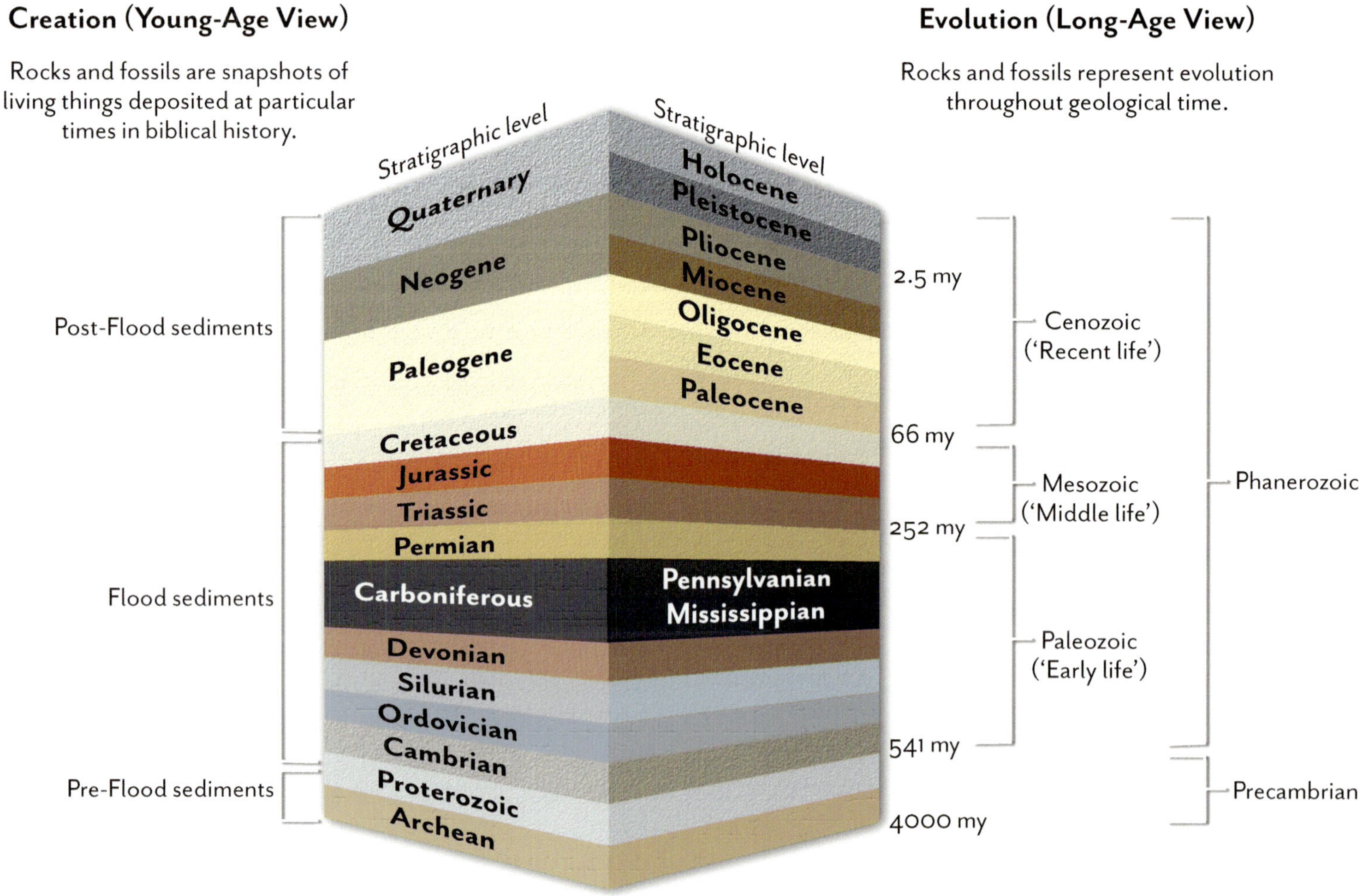

According to evolution, all these creatures shared a common ancestor in the remote past with even simpler animals, plants, and single-celled creatures.

This means that scientists who accept evolution have a very different way of understanding the fossil record than those who accept creation. Evolutionary scientists think that the order in which the fossils appear reflects the order in which living things evolved over hundreds of millions of years.

▼ Evolutionary scientists think that all living things are interrelated and arose from a single common ancestor. This view can be depicted as a single evolutionary tree, and every organism that has ever lived has a place on it somewhere. Note that the creatures in this diagram are not drawn to scale.

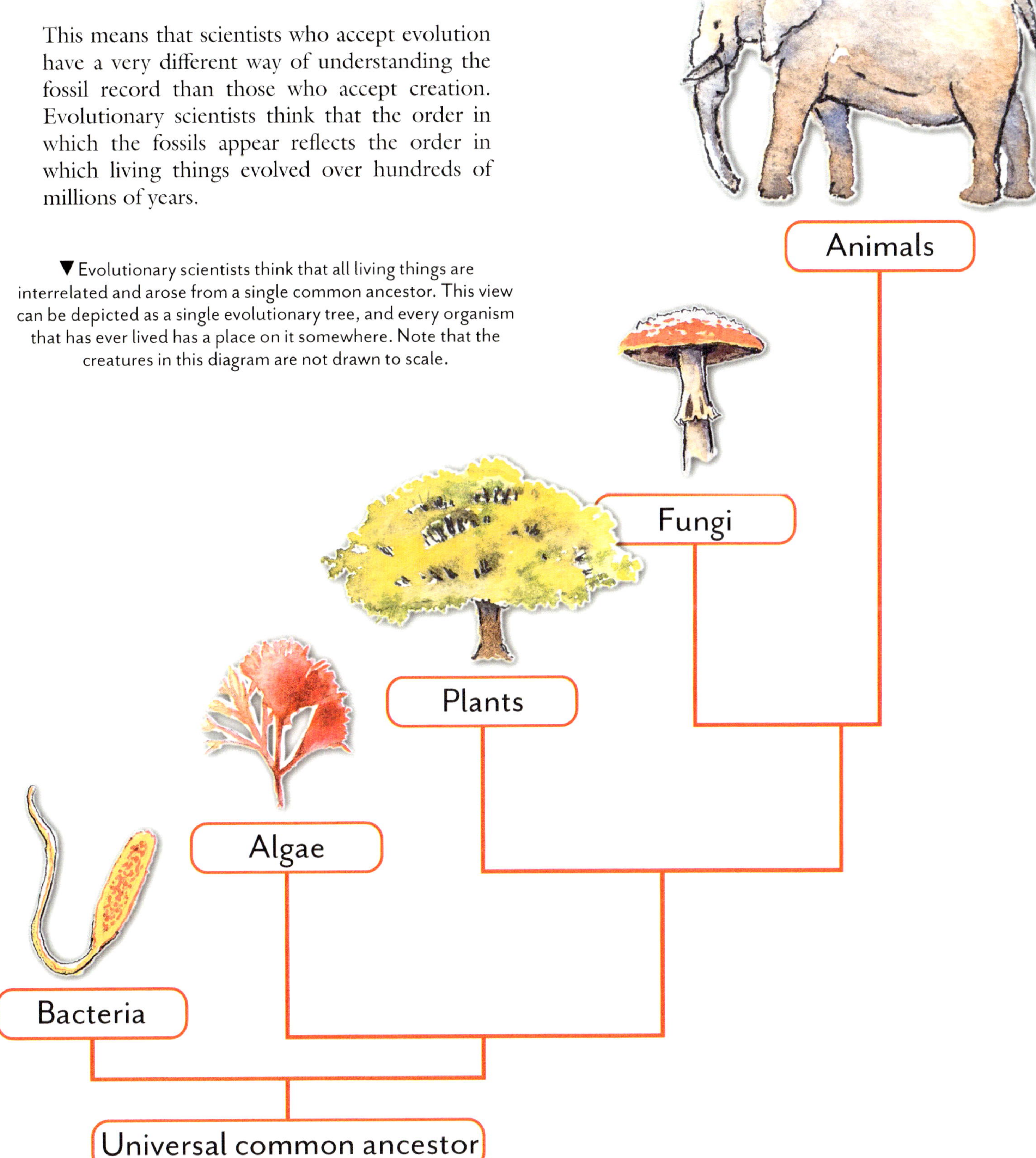

The Bible, however, does not allow for the gradual evolution of life over long ages. According to the Bible, God made everything in six days only thousands of years ago.

Moreover, he made different kinds of creatures on different days of the Creation week—plants on the third day, sea creatures and flying creatures on the fifth day, land creatures and people on the sixth day. They did not evolve from a common ancestor over hundreds of millions of years.

Creationist scientists think that many basic types were separately created in the beginning and that diversification has taken place within each of these groups. This view can be depicted as an orchard of trees, each tree representing a distinctly different group of organisms created by God. Here are three of these created kinds: ▼Grasses (made on Day 3), ▶woodpeckers (made on Day 5), and ◀bears (made on Day 6). Note that the creatures in these diagrams are not drawn to scale.

DID YOU KNOW?

The basic types of organisms that God made in the beginning are sometimes referred to as **baramins**—from the Hebrew words *bara* (meaning *created*) and *min* (meaning *kind*). The creationist biologist Frank Lewis Marsh (1899-1992) introduced the concept of the baramin in 1941, and it has since given rise to an entirely new discipline called **baraminology** for identifying and classifying the created kinds.

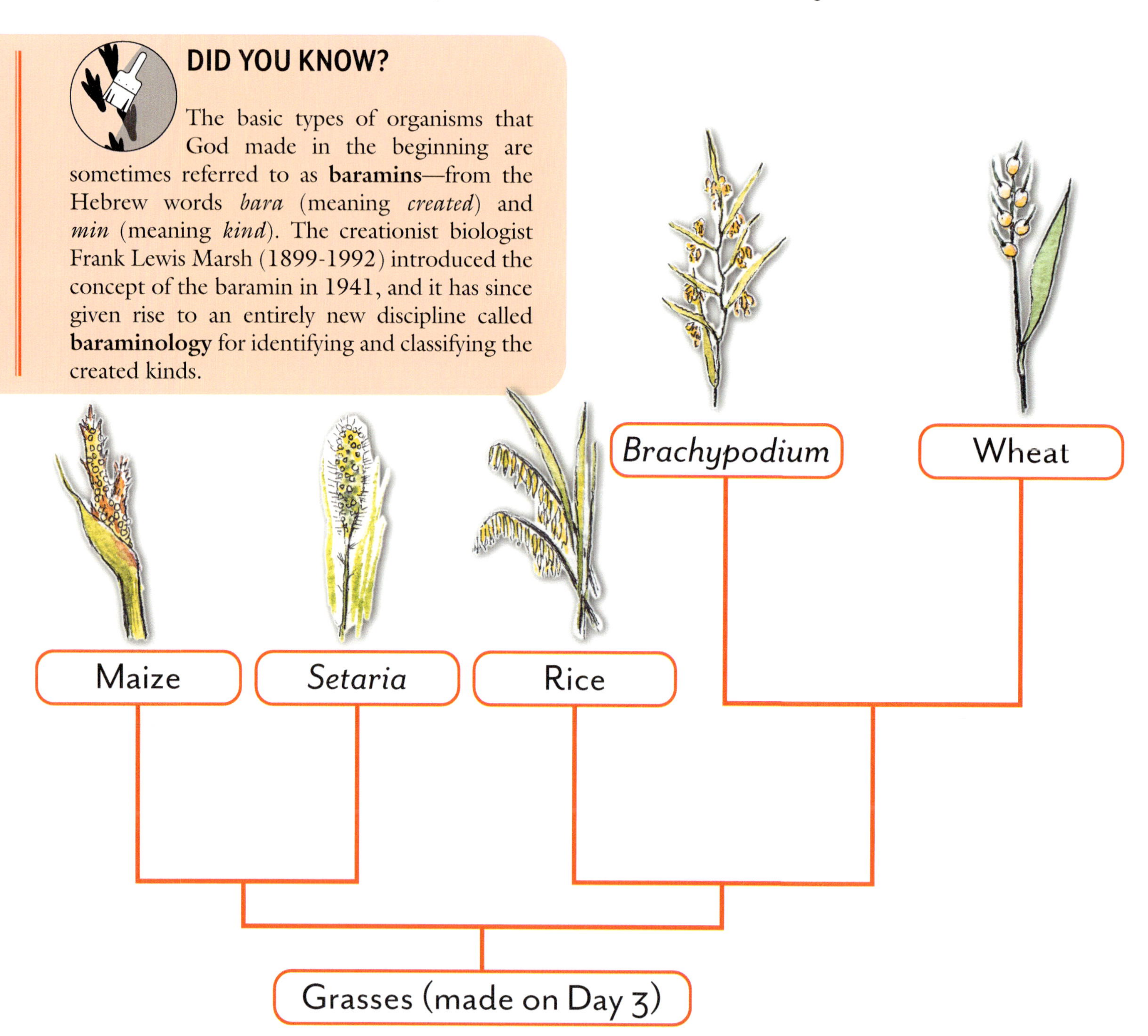

Jynx
Picumnus
Nesoctites
Picus
Campephilus
Woodpeckers (made on Day 5)

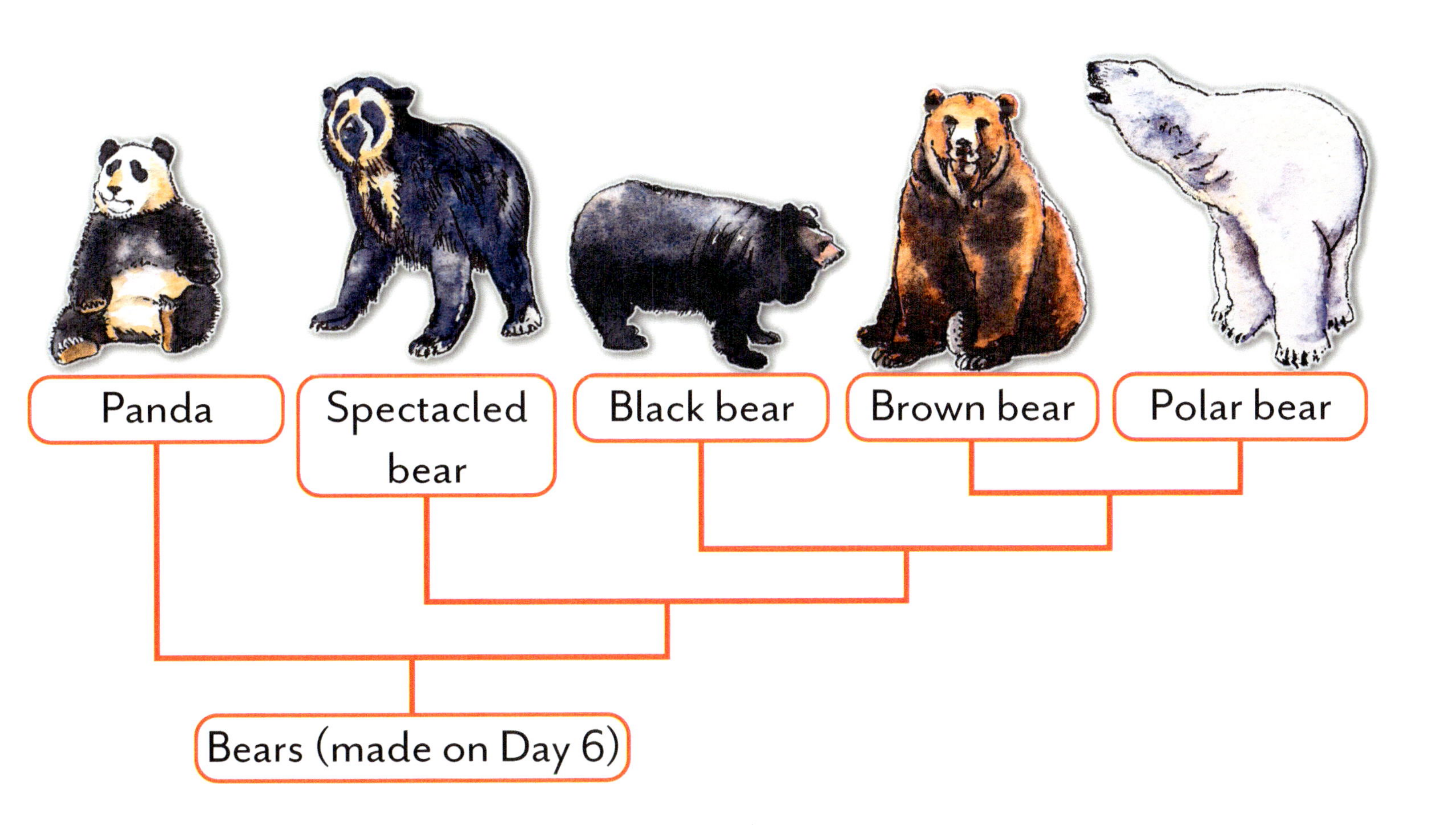
Panda
Spectacled bear
Black bear
Brown bear
Polar bear
Bears (made on Day 6)

DOES EVOLUTION EXPLAIN THE FOSSIL ORDER?

Evolutionary scientists study the shared features that groups of creatures are thought to have inherited from a common ancestor. The scientists' goal is to try to work out how those groups are related. From the pattern of shared traits, they determine the order in which they think the groups split from one another and, therefore, the order in which they would be expected to appear in the fossil record. But in most cases there is a poor match between the actual order of appearance of the fossil groups and the order in which evolutionary theory predicts they should appear. The order of the fossils does not seem to fit the expected evolutionary pattern.

A Predicted

B Observed

▲A hypothetical example illustrating the poor match between most evolutionary trees and the fossil record: A Predicted order of appearance (based on presumed evolutionary relationships). B Actual order of appearance (based on the observed fossil record).

HOW ARE FOSSILS CLASSIFIED?

The scientific discipline concerned with classification is called **taxonomy**, and it is one of the most important branches of biology. The classification of organisms underpins everything that biologists do.

The system we use today was introduced in 1758 by the Swedish biologist and creationist Carl Linnaeus. In this system, animals and plants are grouped into species. Similar species are grouped into genera (singular genus), genera into families, families into orders, orders into classes, classes into phyla, and phyla into kingdoms. Thus, species are best understood as the smallest (least inclusive), stable, recognizable groups with a distinct form (or **morphology**).

Each of these groups is given a name, but creatures are usually identified using the genus and species names (or the genus name alone). For example, the African lion is *Panthera leo* and the domestic dog is *Canis familiaris*. Note that the genus name has an upper-case letter while the species name has a lower-case letter. Both genus and species names are written in italics.

Fossil species are classified in a similar way, but usually we have less information to go on. For example, we do not often have much information about the soft tissues when studying fossilized organisms.

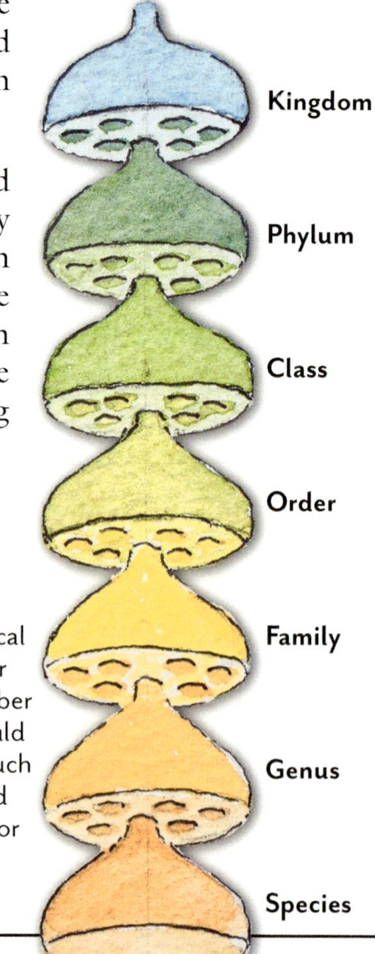

▶The **nested hierarchy** of biological classification showing seven major ranks, or levels. In terms of the number of individuals in each group, it should be noted that lower-level groups (such as species or genera) are small and higher-level groups (such as phyla or kingdoms) are large.

THE FOSSIL TREE WITH MANY NAMES

Sometimes fossils represent parts of a single plant or animal and not the whole creature. These parts may even have been given different names. This fossilized tree has different names for its leaves, cones, spores, stem, and roots.

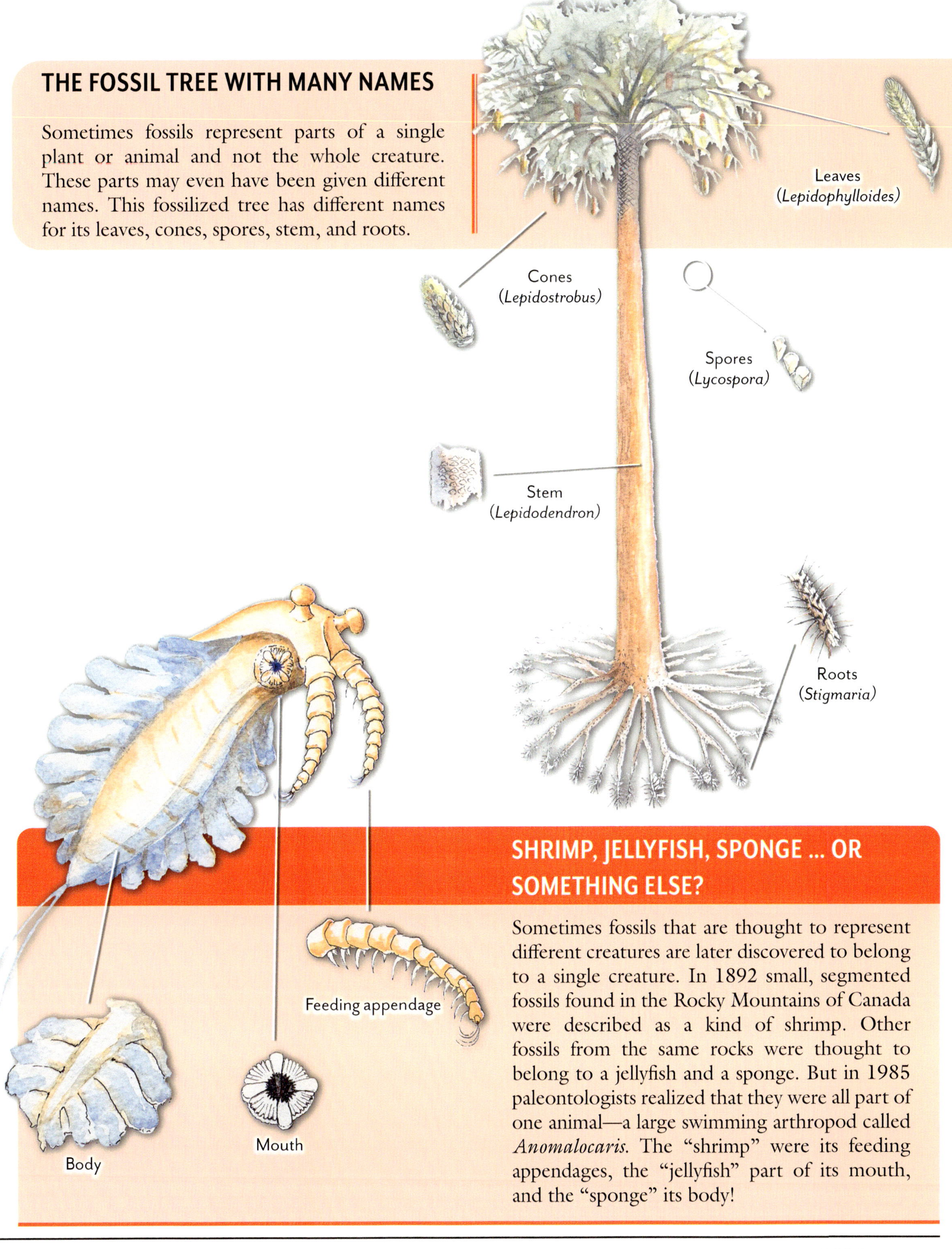

SHRIMP, JELLYFISH, SPONGE ... OR SOMETHING ELSE?

Sometimes fossils that are thought to represent different creatures are later discovered to belong to a single creature. In 1892 small, segmented fossils found in the Rocky Mountains of Canada were described as a kind of shrimp. Other fossils from the same rocks were thought to belong to a jellyfish and a sponge. But in 1985 paleontologists realized that they were all part of one animal—a large swimming arthropod called *Anomalocaris*. The "shrimp" were its feeding appendages, the "jellyfish" part of its mouth, and the "sponge" its body!

MAJOR FOSSIL GROUPS

Among the major groups represented in the fossil record are various kinds of microscopic organisms and plants, invertebrates (animals without backbones), and vertebrates (animals with backbones). These groups are represented by many fossil species, and they come in a spectacular array of shapes, sizes, and designs—every one of them a testimony to God's creative genius.

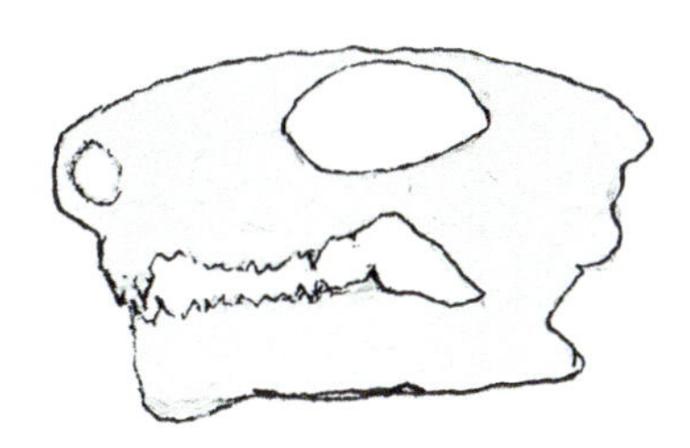
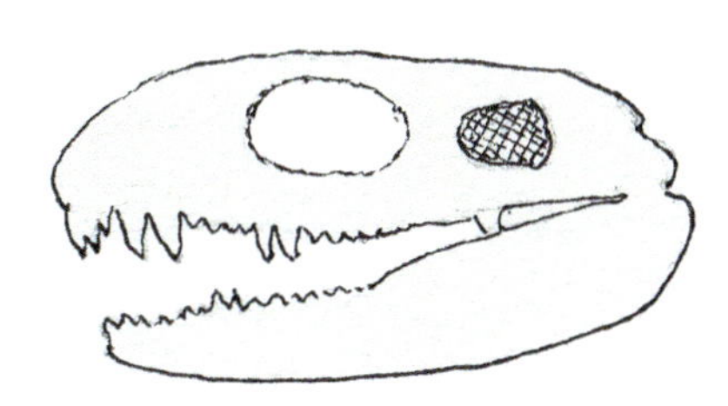
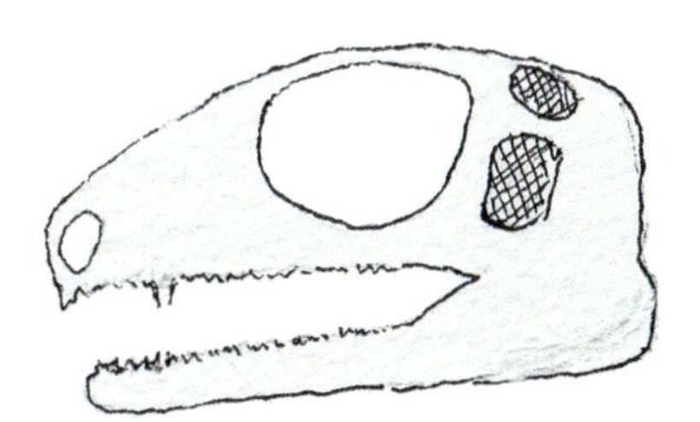

MICROFOSSILS

Microfossils are typically less than one millimeter in size and are usually studied using a microscope. Most are the remains of single-celled creatures, but some are parts of larger organisms.

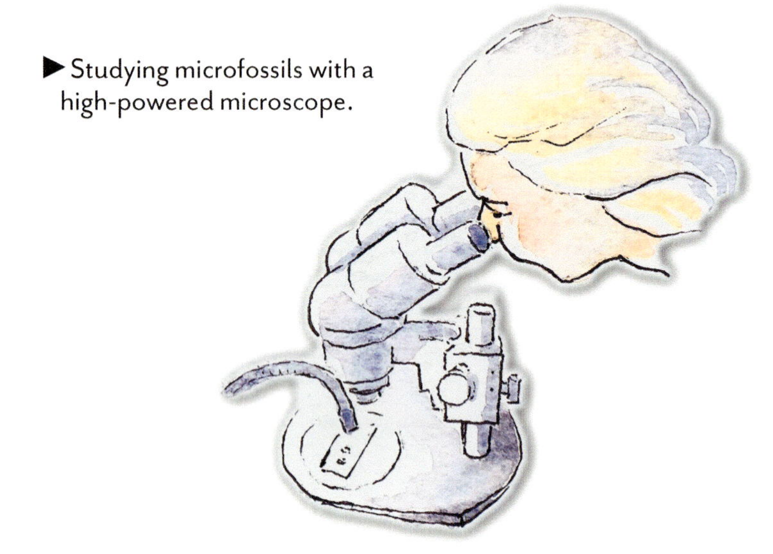
▶ Studying microfossils with a high-powered microscope.

One major group of microfossils are the **cyanobacteria**, which are found in Precambrian to Holocene rocks. Some cyanobacteria form layered structures called **stromatolites**. Stromatolites develop when the microbes form sticky mats on which sediment particles (such as sand grains) get trapped. Over time many layers build up, one on top of another, until larger structures form.

Another group of microfossils are the **foraminifera** (or forams)—amoeba-like creatures with chambered shells. Some make a shell by sticking together silt and sand grains but most secrete a shell of calcium carbonate. Modern forams live in ocean habitats ranging from the deep sea to the beach. Some float in the water (**planktonic** forms) while others live on the sea bottom (**benthic** forms). Their fossils are found in sediments from the Cambrian to the Holocene and are often abundant. Large areas of the ocean floor are covered with soupy sediments made up of foram shells.

▲ Ⓐ Planktonic foraminifera live in the water column while Ⓑ benthic foraminifera live on or in the sediment on the seafloor.

SOME TYPES OF STROMATOLITES

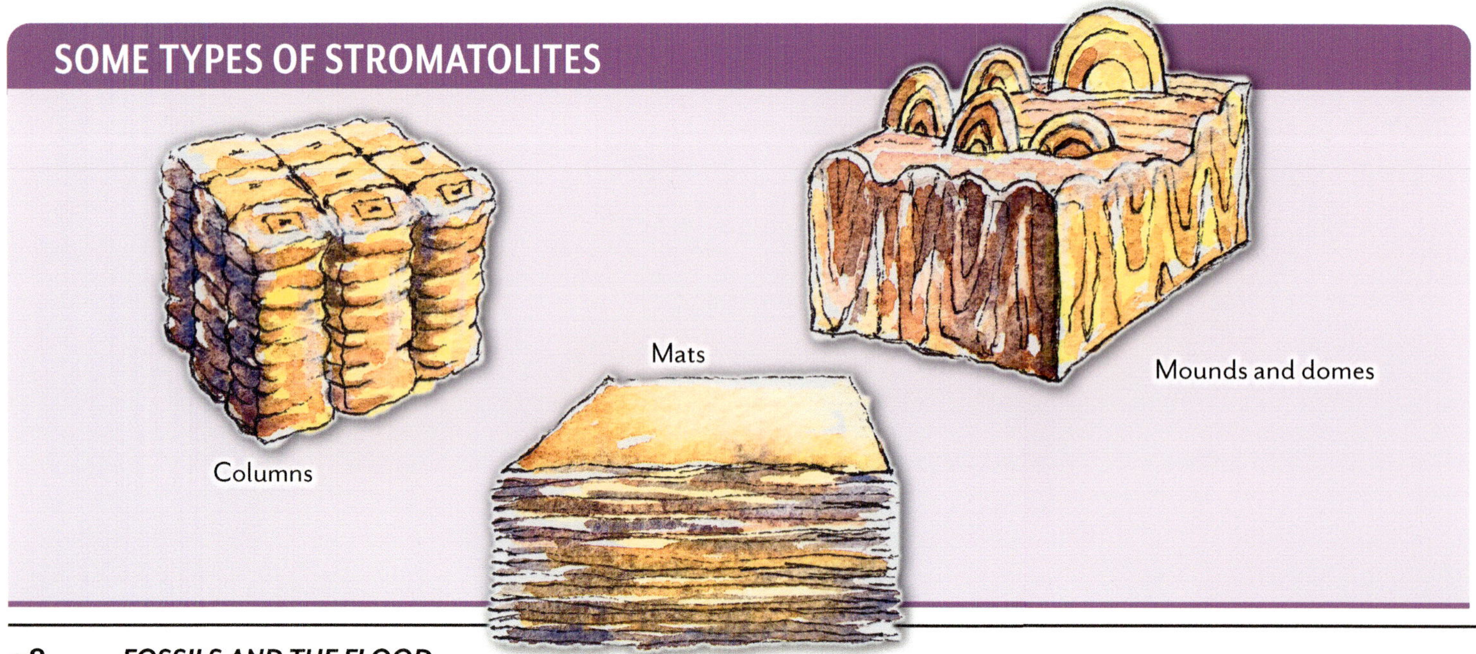

GIANT-SIZED FORAMS

Most forams are about the size of a pinhead but some are much larger.

The extinct **fusulinids** were usually about the size and shape of a wheat grain. Some grew to giant size—up to 2 inches (5 centimeters) long!

Another group of large forams were the lens-shaped **nummulitids**. The first person to write about them was the Greek historian Herodotus (c. 484-425 BC) who saw their fossils in the blocks of limestone used to build the pyramids of ancient Egypt.

▲ A *Parafusulina* is found in Permian rocks and was one of the largest fusulinids. B *Nummulites* is common in Eocene limestones in Egypt. The name comes from the Latin *nummulus* meaning *little coin*.

Radiolaria are single-celled marine creatures with glassy shells made of silica. They are typically smaller than foraminifera. Some are solitary while others form colonies. They float freely in the oceans, and have been found living at depths of up to 13,100 feet (4,000 meters). The fossilized remains of radiolaria are found in sediments ranging from the Cambrian to the Holocene, and their skeletons build up on the modern ocean floor as silica-rich oozes.

▶ The silica shells of radiolarians come in many intricate and beautiful shapes and designs.

Diatoms are single-celled algae with shells made of silica. Each shell has two valves, one slightly smaller than the other so that they fit together. Most diatoms live singly, although some form colonies. Modern diatoms are widespread and can be found living in oceans, lakes, rivers, and soils. Their fossils occur in sediments from the Jurassic to the Holocene. Sediments called diatomites are made up of large quantities of diatom remains and are mined for use as abrasives and filler materials.

▲ Diatoms are enclosed in a shell with two valves, one slightly larger than the other. Some are shaped like a hatbox. Others are shaped like ribbons or stars.

Coccolithophores are single-celled marine algae with calcium carbonate skeletons. They float in the surface waters of the oceans and harvest light from the sun. When conditions are right they can become very abundant, even turning seawater milky white. Their fossilized skeletons are one of the main components of chalky sediments.

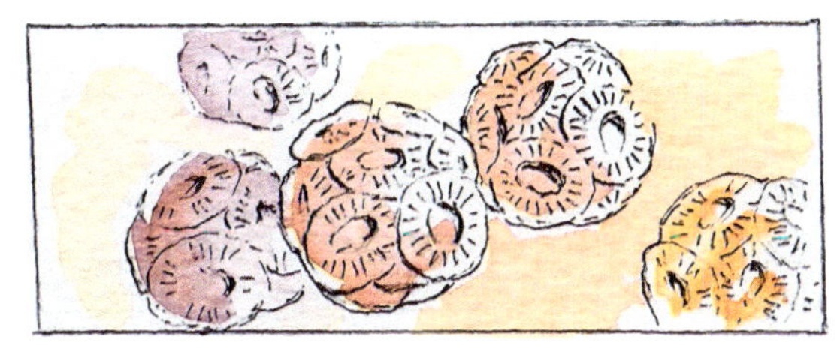

▲ Coccolithophores have shells made of many calcium carbonate plates called coccoliths.

PLANTS

Plants are a diverse group including liverworts, hornworts, and mosses, as well as vascular plants with internal vessels for carrying water and minerals. Some plants are rarely found as fossils while others are more commonly preserved.

Small plants with branching stems but no leaves, roots, or seeds occur in Silurian and Devonian rocks. They had a tough outer layer and a vascular system. Scientists used to call all these plants **psilopsids** but now recognize that they belong to several different groups.

Lycopsids are common plant fossils in the Carboniferous coal deposits. This group includes clubmosses and scale trees. Some were small, but others such as *Sigillaria* and *Lepidodendron* had very tall trunks reaching up to 100 feet (30 meters). Leaves sprouted from their trunks but fell off as the plant grew, leaving distinctive diamond-shaped scars that can still be seen in their fossilized remains.

► A fossilized sheet of *Sigillaria* bark showing its distinctive snake-skin appearance.

▲ Fossil specimens and life reconstructions of three "psilopsids": (A) *Cooksonia* (Silurian and Devonian), (B) *Zosterophyllum* (Silurian and Devonian), and (C) *Psilophyton* (Devonian).

Ferns are common fossils from the Devonian upwards. They reproduce by spores that develop on the undersides of their fronds. **Tree ferns** have a "trunk" of tangled roots with large fronds at the top.

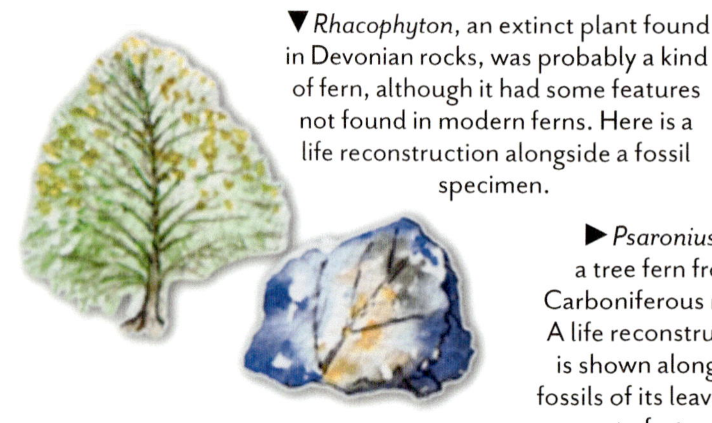

▼ *Rhacophyton*, an extinct plant found in Devonian rocks, was probably a kind of fern, although it had some features not found in modern ferns. Here is a life reconstruction alongside a fossil specimen.

► *Psaronius* is a tree fern from Carboniferous rocks. A life reconstruction is shown alongside fossils of its leaves and part of a trunk.

▼ *Calamites* is a horsetail from Carboniferous rocks. Fossil specimens show the jointed stems alongside a life reconstruction.

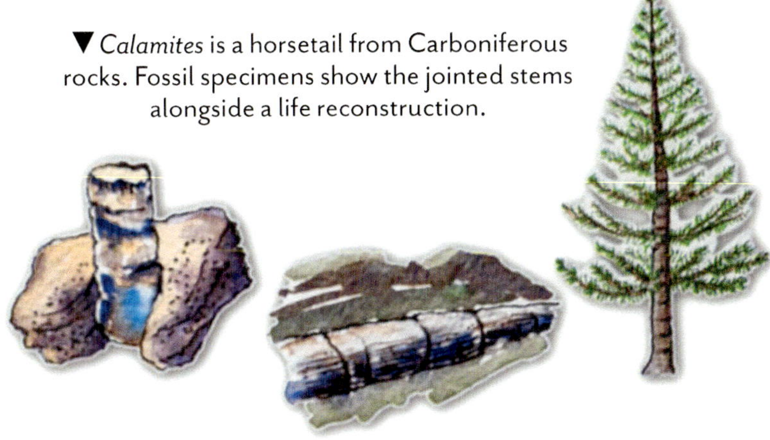

Also common in the Carboniferous coal deposits are horsetails (**sphenopsids**). These plants had jointed stems with many smaller stems or leaves radiating from each joint. Some fossilized horsetails, such as *Calamites*, reached 66 feet (20 meters) tall, quite different from the modern horsetail, *Equisetum*, which is a small rush-like plant that grows around lakes and ponds.

Plants with unprotected seeds (**gymnosperms**) are found in the fossil record from the Carboniferous upwards. These include the **seed ferns**, which resemble true ferns but reproduce by seeds instead of spores; **cycads**, which are palm-like plants that produce seeds inside cones; and **ginkgos**, which have distinctively shaped leaves that are shed annually.

DID YOU KNOW?

Today there is only one surviving ginkgo species, the maidenhair tree (*Ginkgo biloba*), but many more species are found in the fossil record.

▼ A Modern ginkgo leaves (*Ginkgo biloba*) alongside the fossil species B *Ginkgo huttoni* (Jurassic), C *Ginkgo adiantoides* (Upper Cretaceous to Miocene), D *Ginkgo gardneri* (Paleocene), and E *Ginkgo dissecta* (Eocene).

Other gymnosperms known from the fossil record include the **Cordaitales**, tall trees with long leaves and cone-like seed clusters; **Bennettitales**, cycad-like plants that bear seeds in structures resembling flowers; and **conifers**, which have needle-like leaves and produce seeds inside cones.

◀ D Fossil specimen and life reconstruction of *Cordaites*, a cordaitalean tree from Carboniferous rocks, showing its long, strap-like leaves. E Fossil specimen and life reconstruction of *Williamsonia*, an extinct bennettitalean tree found in Lower Jurassic to Upper Cretaceous rocks. F Fossilized cones of a conifer tree (*Araucaria*) from Jurassic sediments in Argentina.

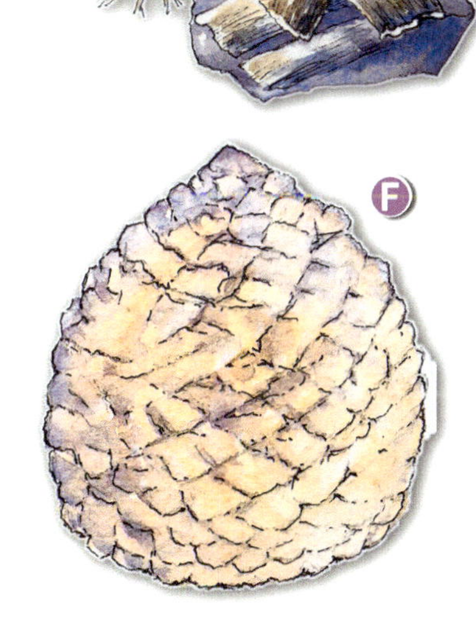

Flowering plants (**angiosperms**) are found in the fossil record from the Jurassic upwards. This very diverse group includes plants such as water lilies, buttercups, magnolias, elms, and poplars, as well as grasses, orchids, and palms. Although fossils of flowering plants are rare, the abundant pollen grains that they produced are often preserved as microfossils in sedimentary rocks.

▶ This rare fossilized flower (*Porana oeningen*) belongs to the morning glory family (the Convolvulaceae) and comes from Upper Miocene rocks in Germany.

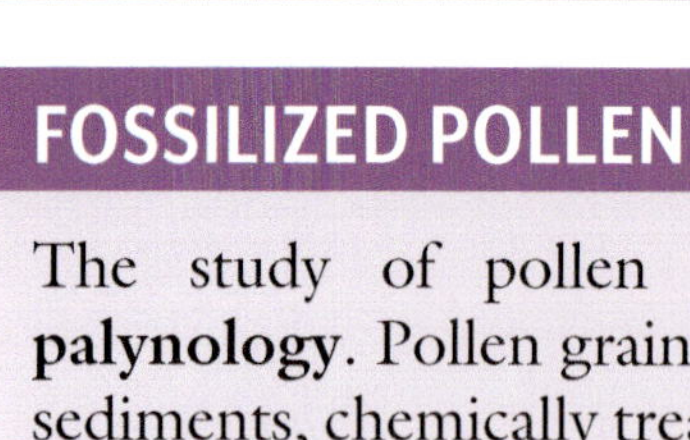

FOSSILIZED POLLEN

The study of pollen and spores is called **palynology**. Pollen grains can be extracted from sediments, chemically treated, and studied under the microscope. Studying pollen allows scientists to work out what vegetation was present in an area in the past, and thus what the climate may have been like.

Pollen grains under the microscope

- A *Pinus* (Pine).
- B *Quercus* (Oak).
- C *Ulmus* (Elm).
- D *Betula* (Birch).
- E *Salix* (Willow).
- F Grass.
- G *Alnus* (Alder).
- H Sedge.

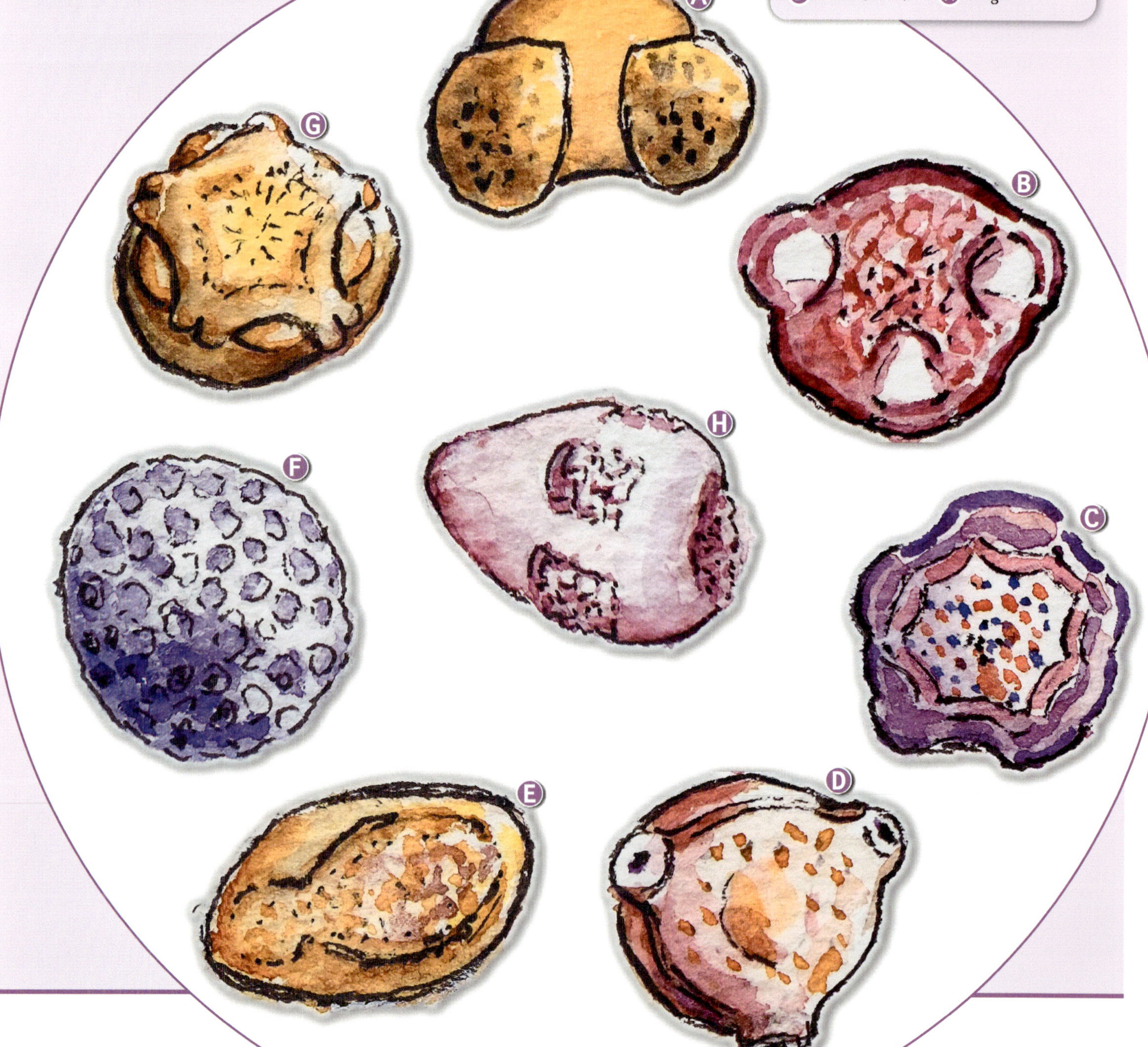

SPONGES, CORALS, AND BRYOZOANS

Many groups of **invertebrates** (animals without backbones) are found in the fossil record.

Sponges are the simplest invertebrates. They have skeletons made of calcium carbonate or silica and they pump water carrying food and oxygen through openings in the skeleton. Fossilized sponges are found in rocks from the Precambrian upwards. Some have distinctive shapes but others can be identified only by careful study with a microscope.

Corals are soft-bodied marine creatures a bit like sea anemones, but with a hard outer skeleton made of calcium carbonate. Some grow singly while others grow in large colonies. They are fixed to the seafloor and waft food into their mouths with their tentacles. Modern corals live in warm, clear, shallow seas and often build large reefs.

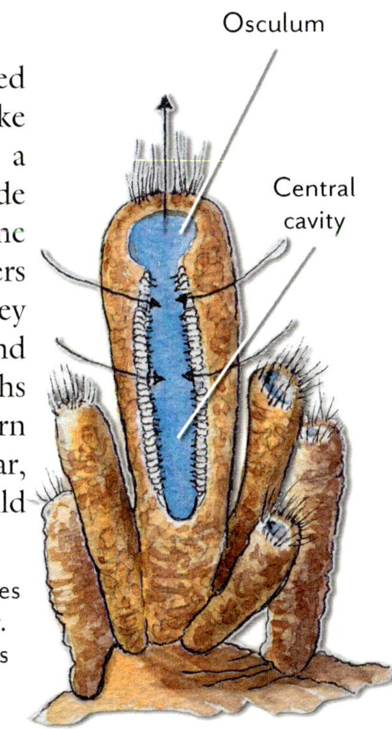

▶ The body of a sponge has thousands of pores that allow water to flow through it constantly. The sponge filters tiny food particles from this flowing water. The arrows in this illustration show the direction of flow.

CORAL REEFS

Coral reefs are structures built by colonies of coral animals. Many modern reefs are found in the warm, clear waters of the tropics, while others occur in cool or cold water.

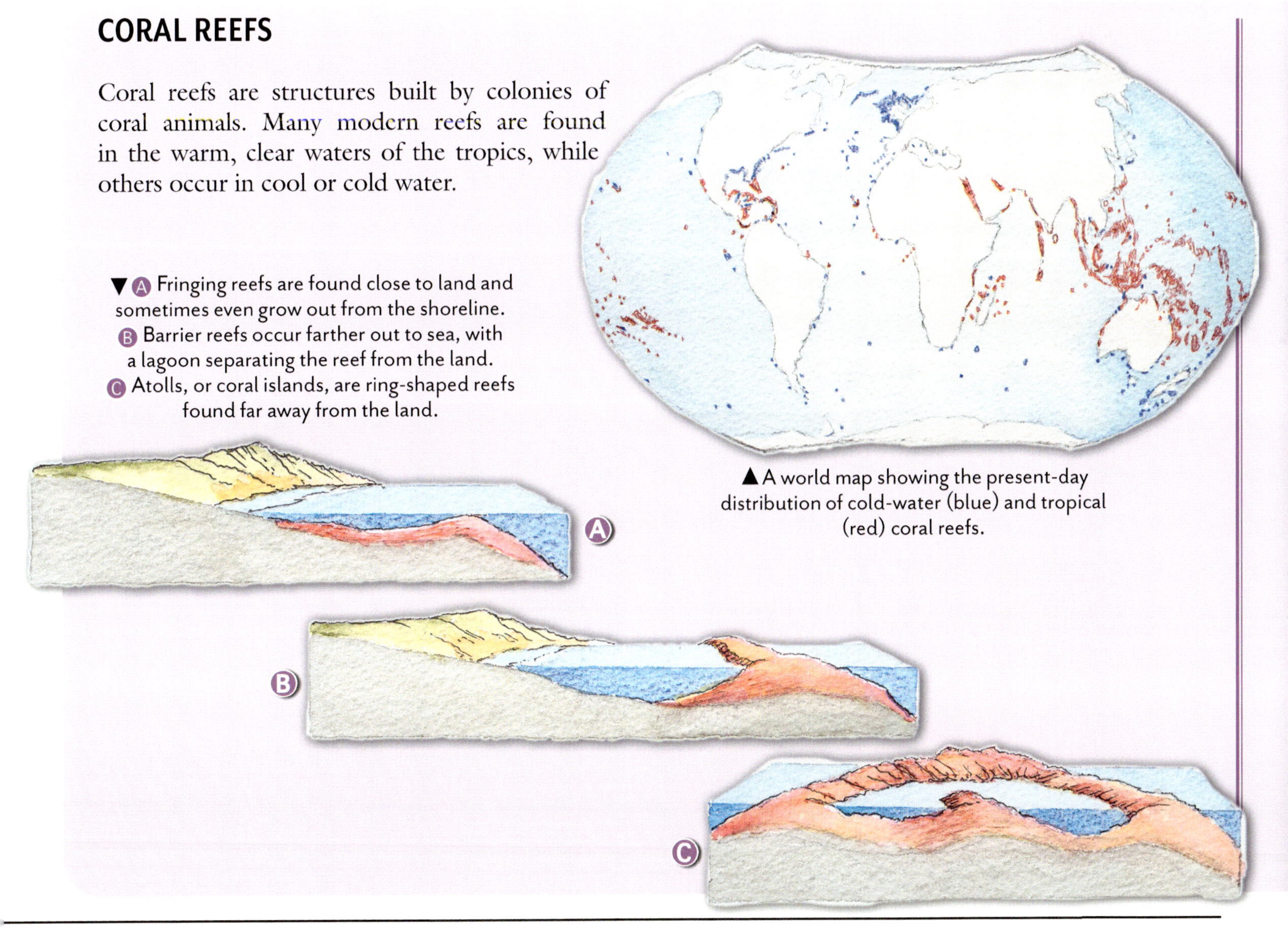

▼ A Fringing reefs are found close to land and sometimes even grow out from the shoreline. B Barrier reefs occur farther out to sea, with a lagoon separating the reef from the land. C Atolls, or coral islands, are ring-shaped reefs found far away from the land.

▲ A world map showing the present-day distribution of cold-water (blue) and tropical (red) coral reefs.

Three groups of corals occur in the fossil record. **Tabulate corals** lived in colonies, while **rugose corals** and **scleractinian corals** included both solitary and colonial forms. Tabulate and rugose corals are found in Ordovician to Permian rocks while scleractinians range from the Triassic to the Holocene.

Bryozoans are invertebrate animals that form lace-like or coral-like colonies. Each member of the colony lives in its own tube with an opening at one end. Most are marine animals, although a few live in freshwater. Fossilized bryozoans are common in many marine limestones from the Ordovician upwards.

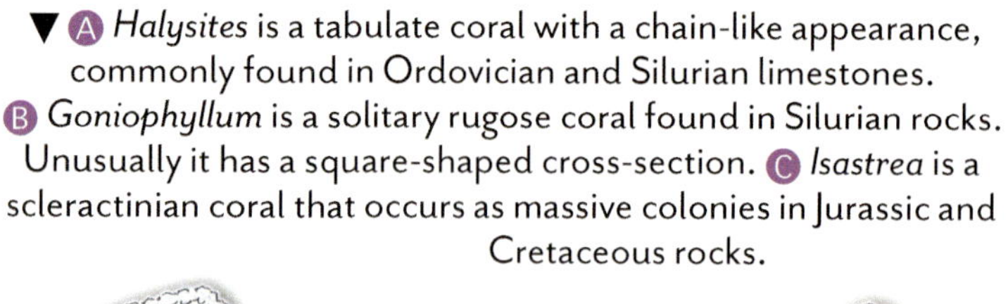

▼ A *Halysites* is a tabulate coral with a chain-like appearance, commonly found in Ordovician and Silurian limestones. B *Goniophyllum* is a solitary rugose coral found in Silurian rocks. Unusually it has a square-shaped cross-section. C *Isastrea* is a scleractinian coral that occurs as massive colonies in Jurassic and Cretaceous rocks.

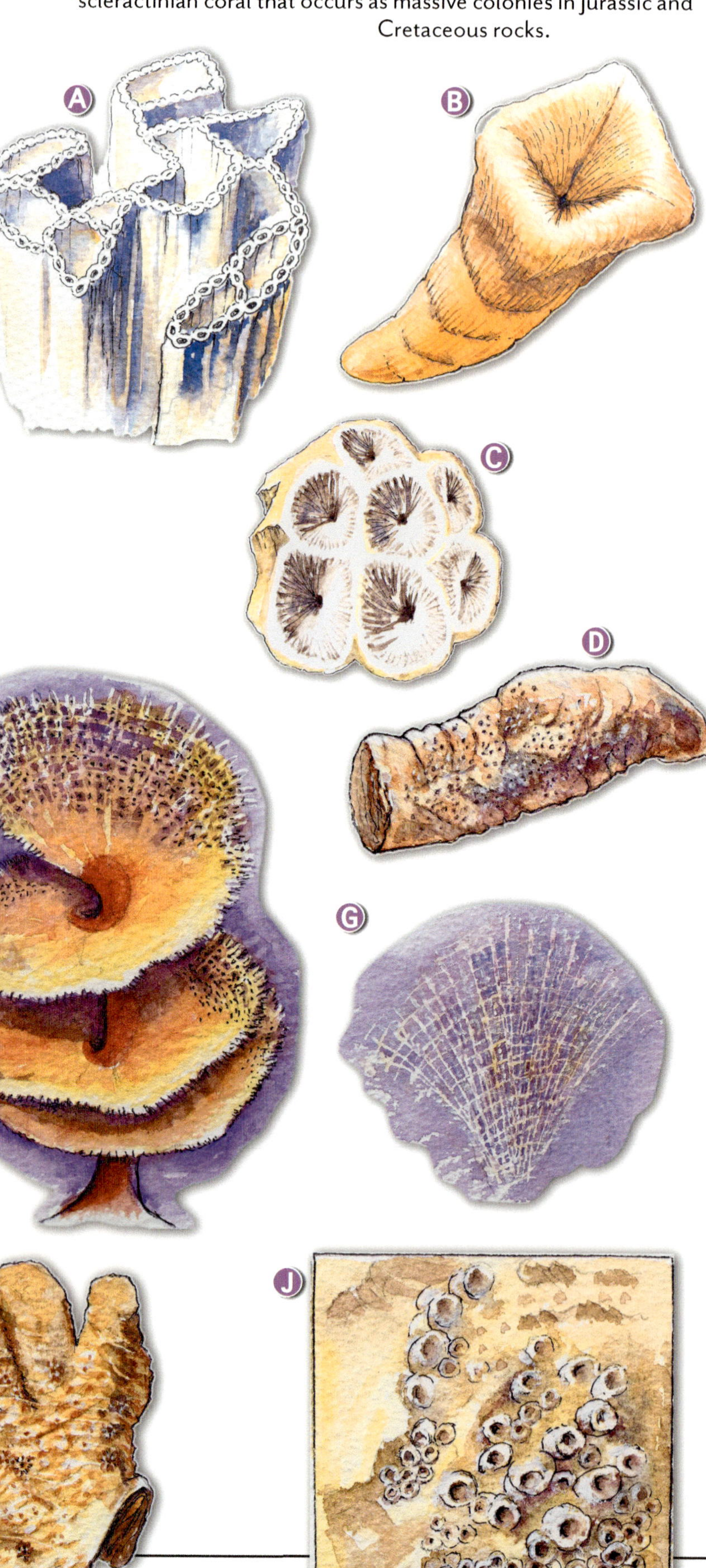

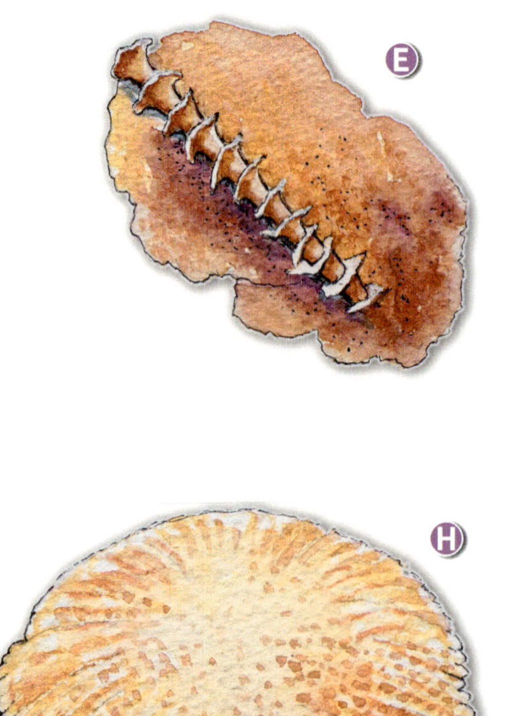

◀ Fossils representing some major bryozoan groups: D *Hexagonella* (a cystoporate). E *Archimedes* (a fenestrate). F Life reconstruction of *Archimedes*. The part usually found fossilized is the central pillar of the animal. The ridges along the pillar supported the delicate coiled netting that housed the individual animals in the colony. G *Fenestella* (a cryptostomate). H *Lunulites* (a cheilostomate). I *Constellaria* (a trepostomate). J *Fistulipora* (a cyclostomate).

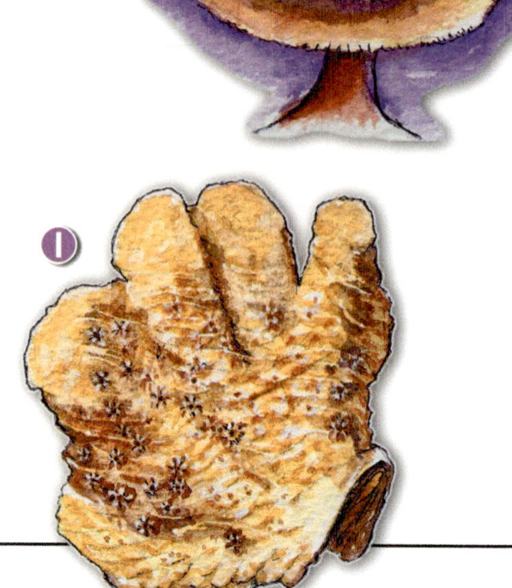

BRACHIOPODS

Brachiopods have shells with two hinged valves. One valve is typically larger than the other. Brachiopods occur in the fossil record from the Cambrian to the Holocene but are especially abundant and diverse in Cambrian to Permian rocks. Today there are fewer than 350 living species. Members of the group are occasionally referred to as lamp shells for their resemblance to ancient oil lamps.

▼ In brachiopods, the valves of the shell cover the top and bottom of the animal. Thus, most brachiopods have symmetrical shells in which the plane of symmetry divides each valve in two. Contrast this with bivalve shells, in which the valves cover the right and left of the animal and the plane of symmetry lies between the two valves (page 107).

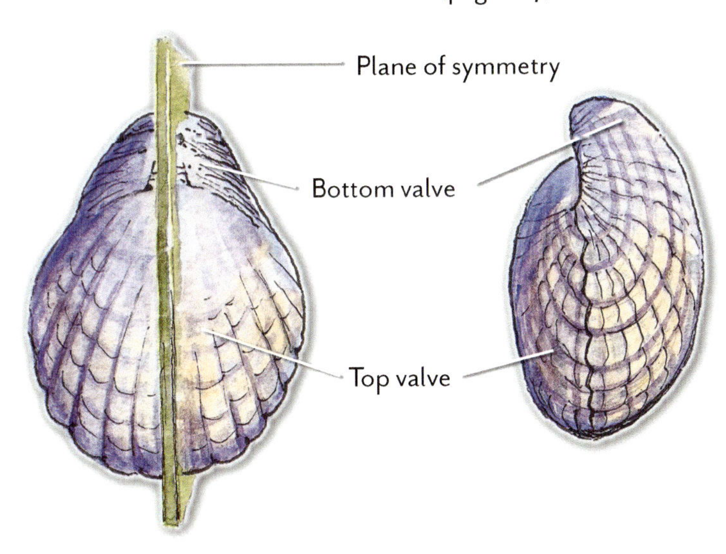

Most brachiopods have shells made of calcium carbonate, but a few have shells made of other types of materials. Most live attached to the seafloor with a fleshy stalk called a **pedicle**. Others burrow into the sediment. They filter food particles from the seawater that passes in and out of their gaping shells.

Broadly speaking there are two types of brachiopods: **articulates** (which have valves joined by a toothed hinge) and **inarticulates** (which have valves without a toothed hinge).

Inarticulate brachiopods are common in Cambrian rocks but less abundant in other rocks. One genus, *Lingula*, burrows into the sediments of mudflats and estuaries around the coasts of modern-day India and Japan. *Lingula* is also represented by fossil species found as far back as the Ordovician.

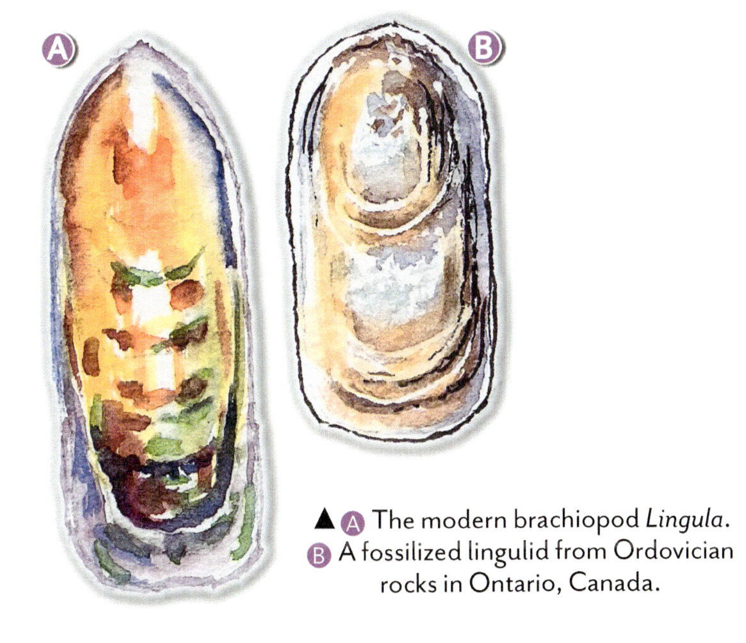

▲ A The modern brachiopod *Lingula*. B A fossilized lingulid from Ordovician rocks in Ontario, Canada.

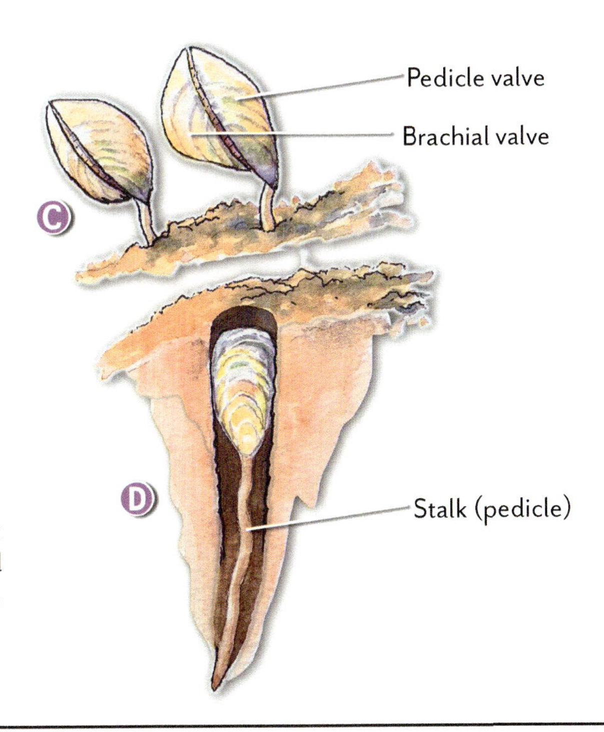

▶ Brachiopods in life position: C attached to the seafloor and D burrowing into the seafloor.

Most fossilized brachiopods are articulates. Major articulate groups include:

Orthids (Cambrian to Permian), which typically have a short hinge line and one valve much flatter than the other.

◀The orthid brachiopod *Dalmanella* is found in Ordovician and Silurian rocks around the world. It has an almost circular outline.

Pentamerids (Cambrian to Devonian), which also have a short hinge line and are often five-sided (pentagonal) in outline.

◀The pentamerid brachiopod *Conchidium* has a shell in which each valve is strongly convex in shape. It is found in Silurian and Devonian rocks worldwide.

Productids (Ordovician to Triassic), which are sometimes spiny and have a long hinge line as wide or almost as wide as the shell.

◀*Gigantoproductus* is one of the largest brachiopods found in Carboniferous rocks, sometimes reaching over 4 inches (10 centimeters) across. It belongs to the group known as productids.

Spiriferids (Ordovician to Jurassic), which usually have a long hinge line and tapering "wings."

◀The spiriferid brachiopod *Spirifer* is common in Carboniferous rocks around the world. It has a characteristically wide shell.

Rhynchonellids (Ordovician to Holocene), which have a short hinge line and cockle-like shells with prominent ribs.

◀The rhynchonellid brachiopod *Goniorhynchia* has a strongly ribbed shell. It is found in the Jurassic rocks of Europe.

Terebratulids (Devonian to Holocene), which typically have a short hinge line and a smooth, bulbous shell.

◀*Gibbithyris* is a terebratulid brachiopod from the Cretaceous rocks of Europe. It has a round shell with a very smooth surface.

DID YOU KNOW?

Spiriferids in the Devonian rocks of Cornwall in southwest England have an unusual, elongated shape, sometimes slightly distorted by earth movements. Their striking appearance gave rise to the local nickname, "the Delabole Butterfly."

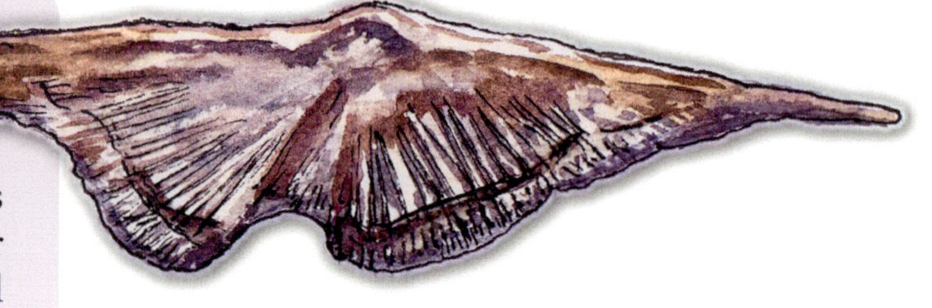

MOLLUSKS I. BIVALVES

There are three groups of **mollusks**: bivalves, gastropods, and cephalopods. Each is well represented in the fossil record. Let us look at each in turn.

Bivalves are found in Cambrian rocks upwards. They have shells with two valves joined by a hinge at the top. Each valve is usually a mirror image of the other, although some bivalves (such as oysters) have one valve larger than the other.

▼ In bivalves, the valves of the shell cover the right and left of the animal. Thus, most bivalves have symmetrical shells in which the plane of symmetry lies between the two valves. Contrast this with brachiopod shells, in which the valves cover the top and bottom of the animal and the plane of symmetry divides each valve in two (page 105).

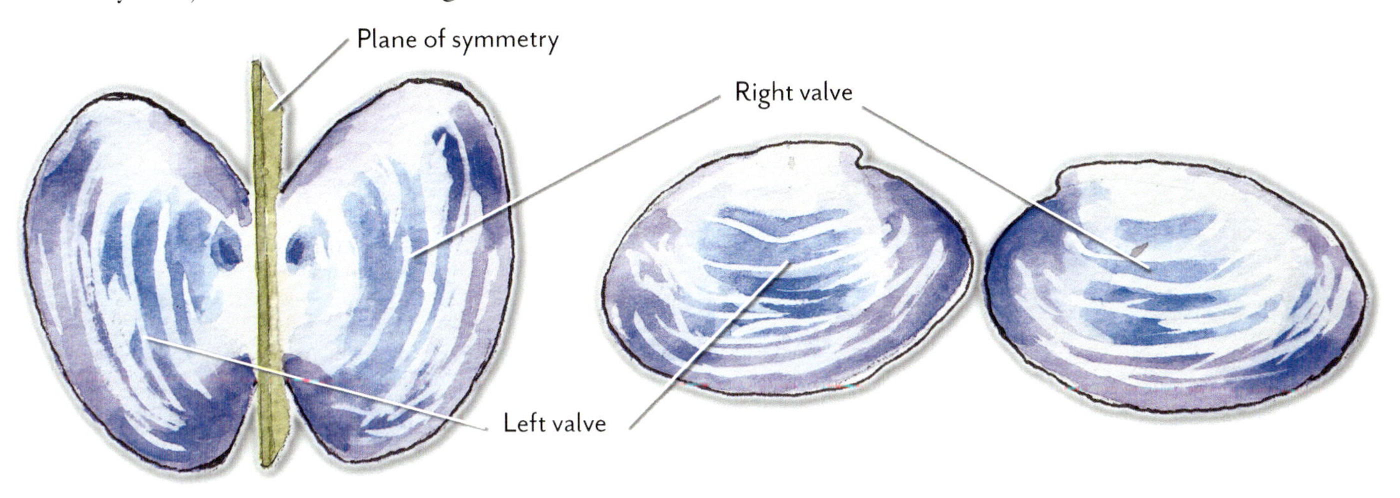

There are four main bivalve groups:

Protobranchs typically have small, nut-shaped shells and obtain food by processing sediments on the seafloor.

▲ *Fordilla* is a protobranch with a small (0.15 inches or 0.4 centimeters) oval shell and is known only from Lower Cambrian rocks.

Anomalodesmatans also burrow into the seabed. But unlike the protobranchs, they tend to have elongated shells and filter their food from seawater.

▲ *Pholadomya* is a Triassic to Holocene anomalodesmatan. It has an elongated shell a few inches long.

Pteriomorphs live on the sediment surface, often in dense clusters and anchored in place by strong filaments. This group includes such familiar types as mussels, oysters, and scallops.

Heterodonts include the cockles, clams, and razor shells. Most species burrow into the seafloor and filter feed, although some bore holes in rocks and other hard surfaces. Perhaps the strangest heterodonts were the extinct rudists, found in Jurassic to Cretaceous sediments. They had horn-shaped shells resembling some solitary corals and even formed reef-like structures.

◀The pteriomorph *Spondylus* is up to 4.75 inches (12 centimeters) long and often quite spiny. It is found across the world in Jurassic to Holocene rocks.

▶ Ⓐ *Glycymeris* is a Cretaceous to Holocene cockle with an almost circular outline. Ⓑ *Hippurites* is a Cretaceous rudist, with a horn-like right valve and a flattened left valve that forms a kind of lid.

A

Outer surface

Inner surface

Left valve

B

Right valve

DID YOU KNOW?

The oyster, *Gryphaea*, which is very common in Jurassic rocks, has a shell shaped a bit like a claw. For this reason *Gryphaea* fossils are sometimes referred to as "the devil's toenails!"

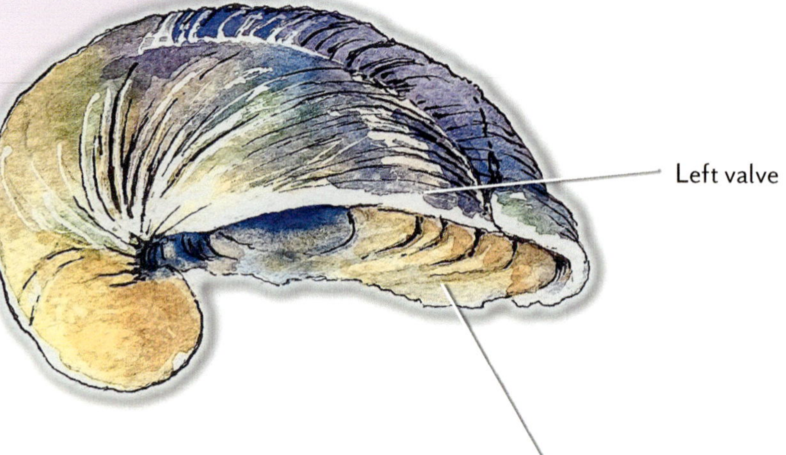

▶*Gryphaea* has a large left valve that curves over onto the right valve, giving it a claw-like appearance.

MOLLUSKS II. GASTROPODS

Gastropods (or snails) are a group of mollusks with about 40,000 living species. They are found as fossils in Cambrian rocks upwards. Most have coiled shells, often raised in the center to form a cone. Some species (such as limpets) are not coiled, and others (such as slugs and sea slugs) have no shells.

There are five main groups of gastropods:

Patellogastropods are the true limpets and have conical shells with a wide opening. Limpets feed on algae, which they scrape from rock surfaces with their rasping mouthparts.

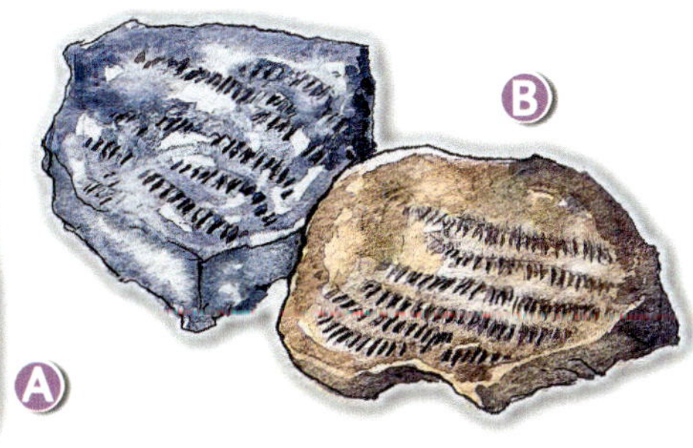

▲ A A grazing trace made by the **radula** (rasping mouth) of a modern limpet. B Similar traces found in the fossil record are named *Radulichnus*.

Vetigastropods are the slit shells, so named because they have coiled shells with a secondary slit-like aperture near the main opening.

▲ The vetigastropod *Discohelix* is a marine form known from Triassic to Oligocene rocks.

Neritimorphs are coiled and typically rather globular in shape. The group includes marine, freshwater, and land species.

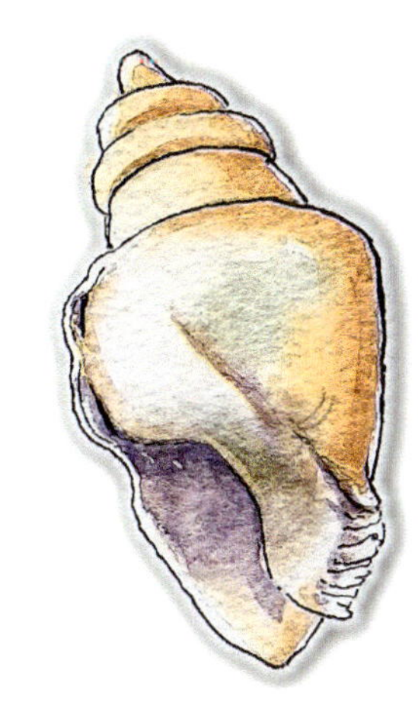

◀ The neritimorph *Melanopsis* is a freshwater gastropod found in Cretaceous to Holocene rocks.

Caenogastropods are very diverse, including periwinkles, whelks, cowries, and many other types. Some species are grazers, while others are scavengers, filter feeders, or carnivores.

▼ An assemblage of the caenogastropod *Turritella*, a high-spired form that is common in some Cretaceous to Holocene sediments.

Heterobranchs include the bubble shells and the high-spired nerinoids. This group also includes the sea slugs, but these animals have no fossil record because they are soft-bodied and do not preserve well.

◀ *Aptyxiella* is a heterobranch gastropod found in the Jurassic rocks of Dorset on England's southern coast. Its appearance gave rise to its local nickname, "the Portland Screw."

MOLLUSKS III. CEPHALOPODS

Cephalopods are mollusks with tentacles surrounding the mouth. Modern cephalopods include shelled forms (such as *Nautilus*) and soft-bodied forms (such as squids and octopuses). Some types (such as cuttlefish) have an internal shell. An even greater range of cephalopods, however, is seen in the fossil record. Most fossilized cephalopods possess a shell with many chambers.

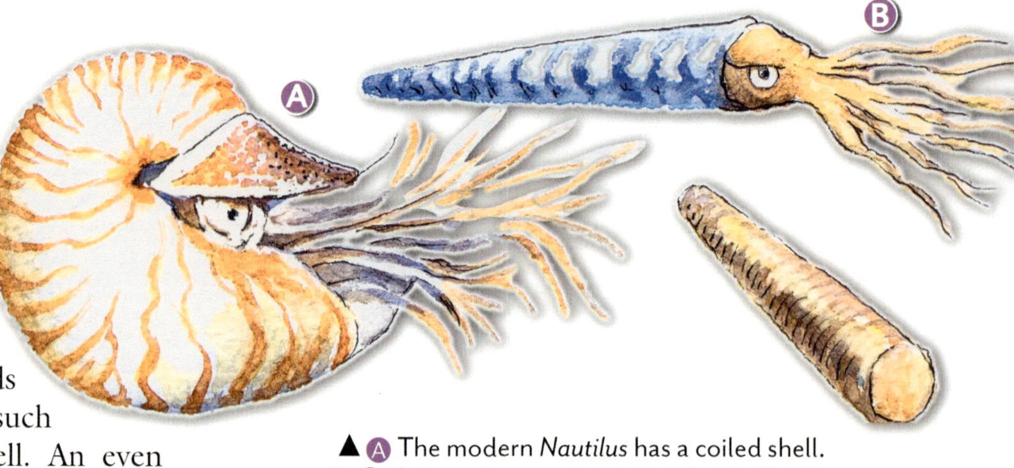

▲ A The modern *Nautilus* has a coiled shell. B *Orthoceras* is an extinct, straight-shelled form known from Ordovician to Triassic rocks, shown here in life reconstruction and in fossil form.

Nautiloids are found in rocks ranging from the Cambrian to the Holocene. Most Cambrian and Ordovician types had straight shells and some reached several yards long. Other fossil species resembled the modern *Nautilus* and had coiled shells.

▲ C *Acanthoceras* is a typical coiled ammonite known from Upper Cretaceous rocks. D *Didymoceras* is an unusual heteromorphic ammonite, also discovered in Upper Cretaceous rocks. A fossil specimen is shown here alongside a life reconstruction.

Ammonoids occur in Devonian to Cretaceous rocks. There were three main groups: the **goniatites**, the **ceratites**, and the true **ammonites**. Ammonites are popular with collectors because their fossils are so varied and abundant. They had shells that were often ornamented and ranged from an inch to several yards across. Some even had uncoiled or partially coiled shells (**heteromorphs**).

Belemnoids range from the Carboniferous to the Eocene. They are particularly common in Jurassic and Cretaceous rocks. Belemnoids were cuttlefish-like animals. Their bullet-shaped fossils are the remains of the internal shell part called the guard or **rostrum**—the equivalent of a modern cuttlefish bone.

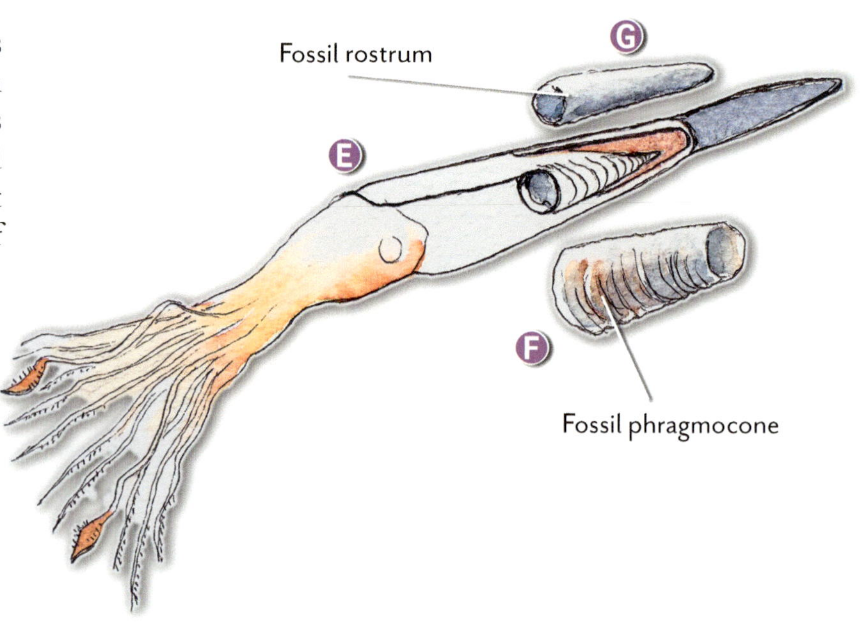

▶ Belemnoids had an internal shell made of three parts: E The **proostracum**, F the **phragmocone**, and G the rostrum. The bullet-shaped rostrum is the part that is most often preserved in the fossil record.

IDENTIFYING FOSSILIZED CEPHALOPODS FROM THEIR SUTURE PATTERNS
Nautiloids and ammonoids display distinctive suture patterns, lines that mark where the partitions separating each chamber met the external shell. These can be used to distinguish the major groups.
Nautiloids: Simple (straight or curved) sutures.
Goniatites: Jagged sutures with saddles (pointing to the front of the animal) and lobes (pointing to the back of the animal).
Ceratites: Sutures with rounded saddles and frilly lobes.
Ammonites: Complex sutures with frilly saddles and lobes.

ARTHROPODS I. TRILOBITES

Arthropods are invertebrates with a hard outer skeleton and jointed legs. This group includes crabs and lobsters, insects and spiders, as well as many creatures that are now extinct.

Trilobites are the most well-represented arthropods in the fossil record. They are common fossils and popular with collectors. Over 15,000 species are known in rocks ranging from the Cambrian to the Permian.

Trilobites had a head (**cephalon**), a body made of segments (**thorax**), and a tail (**pygidium**). Each body segment had a pair of legs. A central ridge ran down the body, dividing it into three lobes, which gives the group its name.

Most adult trilobites were about an inch (2.5 centimeters) long. Some, though, reached only a quarter of an inch (0.6 centimeters) (including *Agnostus*, a tiny, blind trilobite). Others grew to over a foot and a half (0.45 meters) (such as *Isotelus*, the largest trilobite so far discovered).

MAJOR TRILOBITE GROUPS

Agnostida (Cambrian to Ordovician)
◀ *Agnostus* from the Cambrian and Ordovician of Europe and Asia.

Redlichiida (Cambrian)
▶ *Paradoxides* from the Cambrian of North America, Europe, Africa, and Australasia.

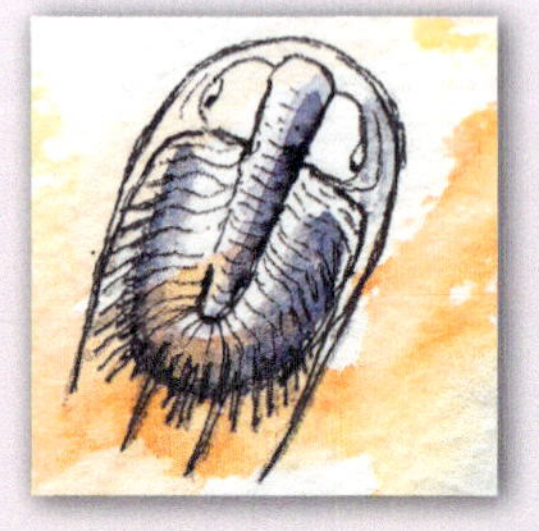

Corynexochida (Cambrian to Devonian)
◀ *Oryctocephalus* from the Cambrian of North America, South America, Europe, and Asia.

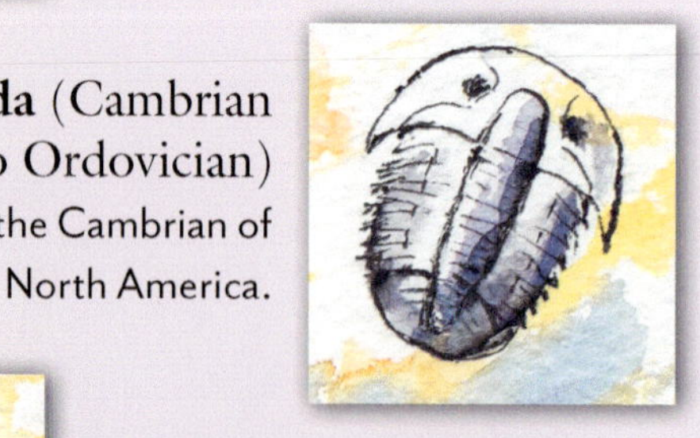

Ptychopariida (Cambrian to Ordovician)
▶ *Elrathia* from the Cambrian of North America.

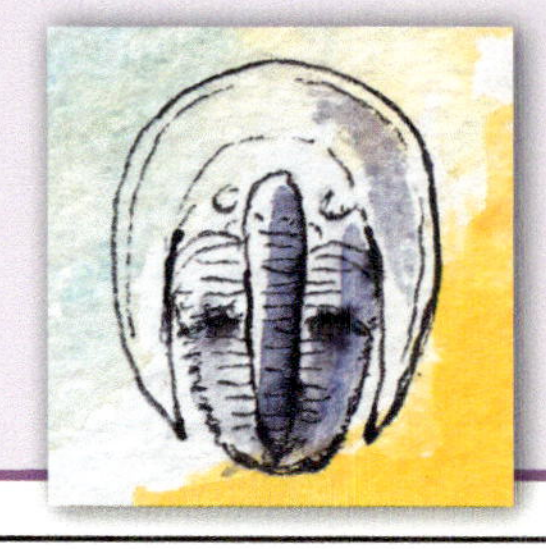

Harpetida (Cambrian to Devonian)
◀ *Harpes* from the Devonian of Europe and Africa.

Proetida (Ordovician to Permian)
◀ *Cyphoproetus* from the Ordovician and Silurian of North America and Europe.

Phacopida (Ordovician to Devonian)
▶ *Phacops* from Silurian and Devonian rocks across the world.

Lichida (Cambrian to Devonian)
◀ *Arctinurus* from the Silurian of North America and Europe.

Asaphida (Cambrian to Silurian)
▶ *Isotelus* from the Ordovician of North America, Europe, and Asia.

TRILOBITE EYES

Trilobites were complex animals. Some species were blind and may have burrowed into the seafloor. Others had extraordinary compound eyes with hundreds of crystal lenses, and were probably active swimmers. These eyes were remarkably designed to produce all-round vision and a well-focused image.

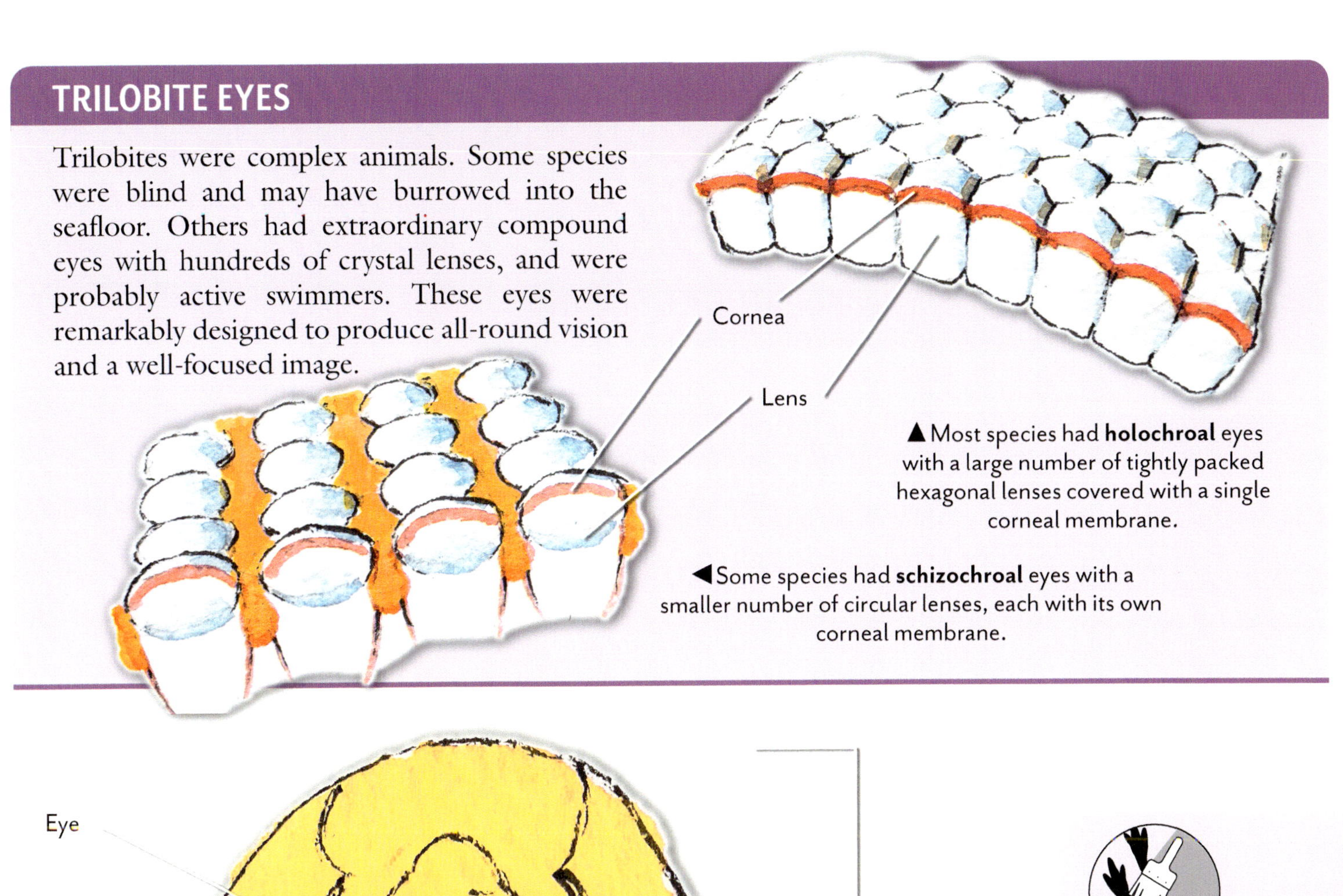

▲ Most species had **holochroal** eyes with a large number of tightly packed hexagonal lenses covered with a single corneal membrane.

◀ Some species had **schizochroal** eyes with a smaller number of circular lenses, each with its own corneal membrane.

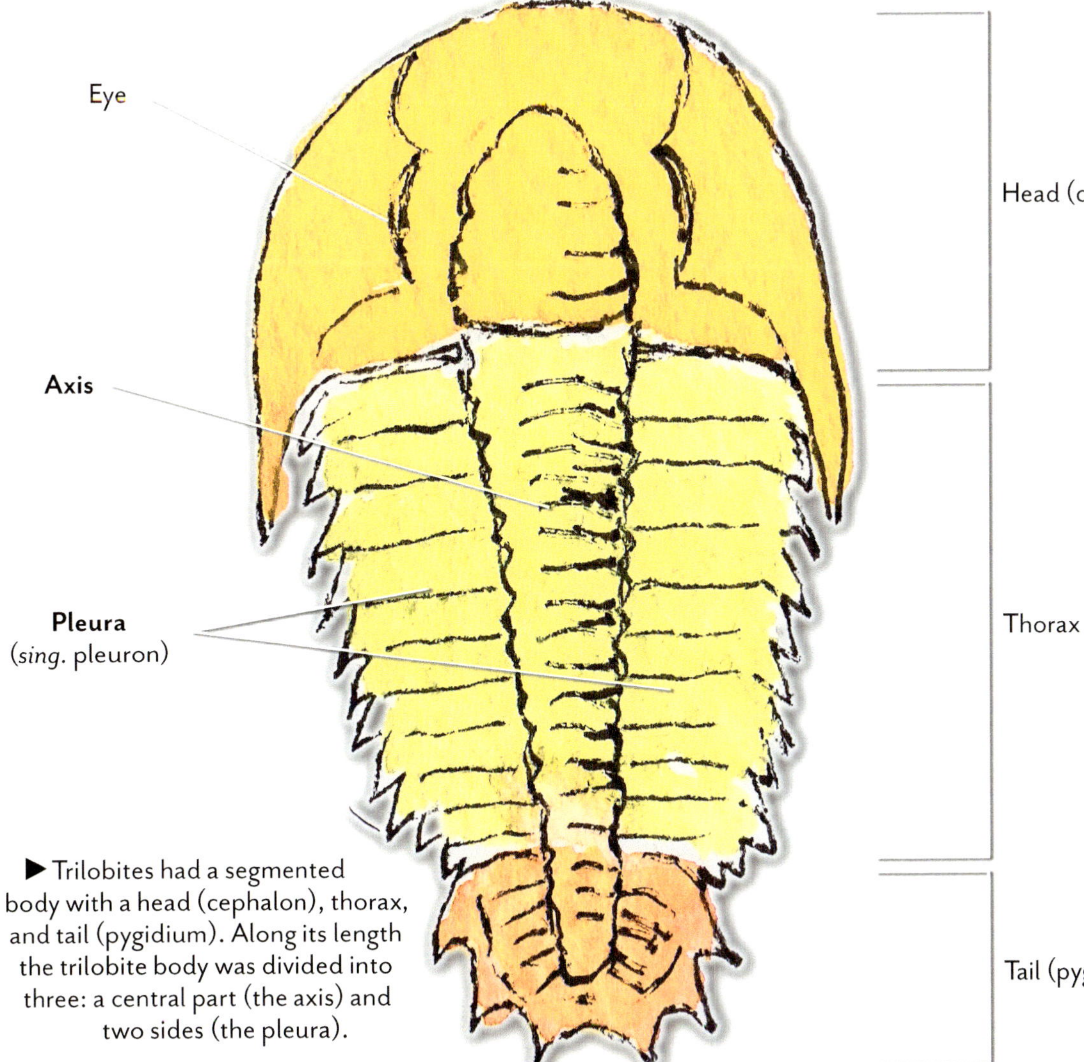

▶ Trilobites had a segmented body with a head (cephalon), thorax, and tail (pygidium). Along its length the trilobite body was divided into three: a central part (the axis) and two sides (the pleura).

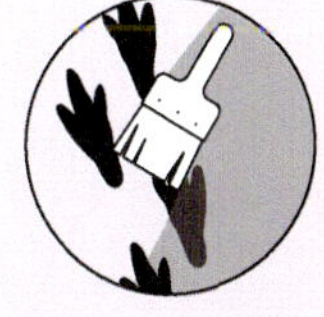

DID YOU KNOW?

The hard outer skeleton of a trilobite could not grow. Instead it would be cast off and regrown many times during the life of an individual trilobite. Many trilobite fossils are thought to be the remains of these cast-offs rather than a complete animal.

ARTHROPODS II. CHELICERATES, CRUSTACEANS, MYRIAPODS, AND INSECTS

▼The extinct eurypterids included some spectacular forms: Ⓐ Fossil specimen of *Pterygotus* from the Silurian and Devonian of Europe and North America. Ⓑ Life reconstruction.

Besides trilobites, many other arthropod groups are represented in the fossil record.

Chelicerates is a group that includes spiders, scorpions, horseshoe crabs, and extinct forms called eurypterids. The group receives its name from the pair of pincers (**chelicerae**) attached to the first body segment of these animals. Fossils of chelicerates are not common but have been found in rocks ranging from the Cambrian to the Holocene. The extinct **eurypterids** are found in Ordovician to Permian rocks and resembled water scorpions. Most species were less than 8 inches (20 centimeters) long, but others, such as *Pterygotus*, were gigantic and reached almost 6.5 feet (2 meters) long.

Crustaceans include marine species (such as crabs) as well as freshwater and land forms (such as wood lice). Their fossilized remains occur in Cambrian to Holocene sediments. The most important fossilized crustaceans are an unusual group called **ostracods**, which have a body enclosed between two bean-shaped valves. Most species are small (less than one millimeter across) but some reach a few centimeters long. They are found living in soils, lakes, and oceans.

Myriapods include centipedes and millipedes. They are found in Silurian to Holocene rocks. In fact, a myriapod called *Arthropleura* was the largest land-dwelling arthropod. Its fossils are found in Carboniferous rocks and it is thought to have reached at least 6.5 feet (2 meters) in length.

▶Some fossilized myriapods were truly gigantic: Ⓒ Two parallel *Arthropleura* trackways exposed along the foreshore near the village of Crail in southeast Fife, Scotland. Ⓓ Life reconstruction of *Arthropleura*, a 6.5-foot-form (2 meters) from Carboniferous rocks in Europe and North America.

Insects include dragonflies, ants, beetles, wasps, and flies. This is a very numerous and diverse group, but its fossil record is fairly poor. Some exceptional fossil beds preserve an impressive range of species, such as those at Solnhofen in Germany (Jurassic) and Florissant in Colorado, USA (Eocene). Fossilized insects are also well preserved in some amber deposits.

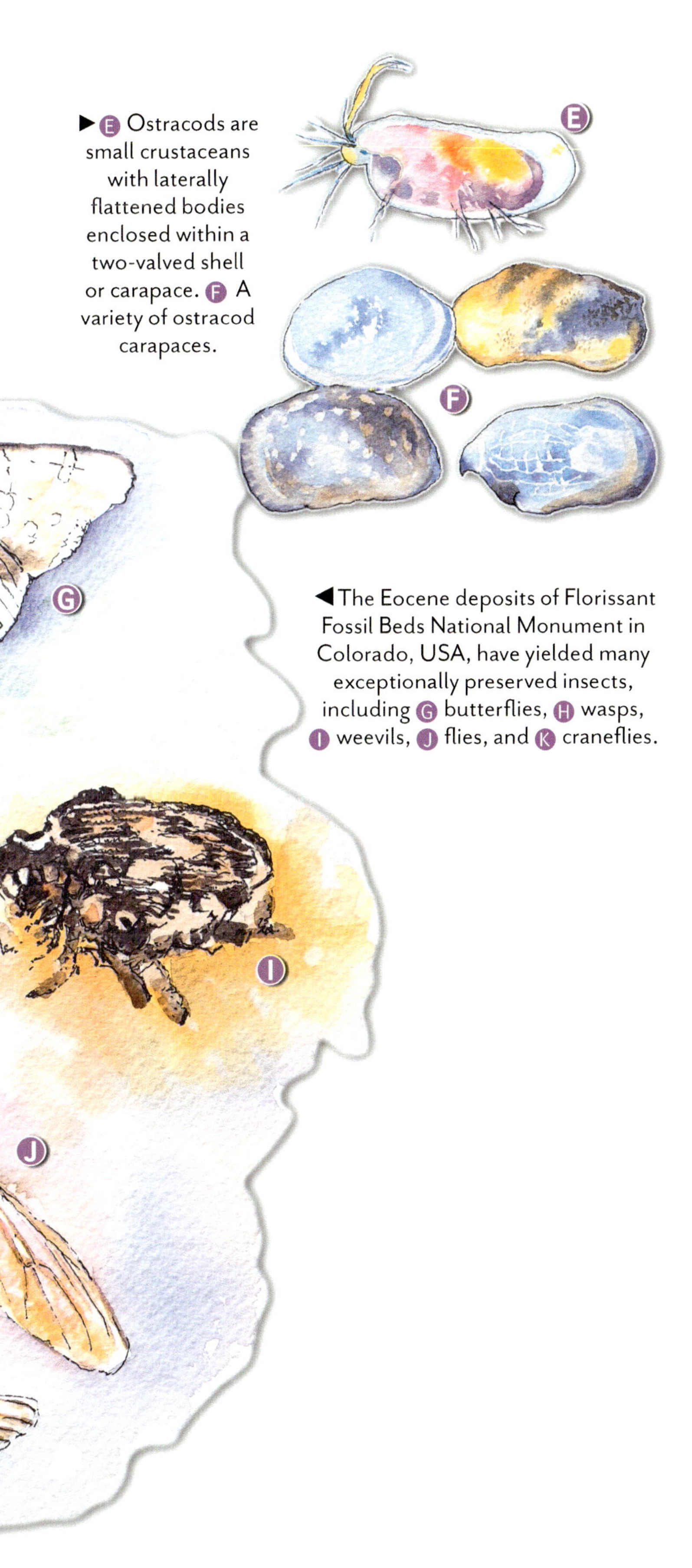

► E Ostracods are small crustaceans with laterally flattened bodies enclosed within a two-valved shell or carapace. F A variety of ostracod carapaces.

◄ The Eocene deposits of Florissant Fossil Beds National Monument in Colorado, USA, have yielded many exceptionally preserved insects, including G butterflies, H wasps, I weevils, J flies, and K craneflies.

ECHINODERMS

Echinoderms (meaning *spiny skins*) include familiar animals such as sea urchins, starfish, and sea lilies, but also some less familiar extinct types. Members of the group have skeletons made of many hard calcium carbonate plates, special tube feet for moving around, and a five-rayed body pattern.

Several groups of echinoderms are represented in the fossil record:

Sea urchins (**echinoids**) are globular animals with spines. Regular forms are usually round and have a mouth at the base and an anus at the top. Irregular forms are more heart-shaped and the anus is nearer the back of the animal.

Sea lilies (**crinoids**) are rather plant-like in appearance with a stem, a body made of many plates, and five or more branching arms.

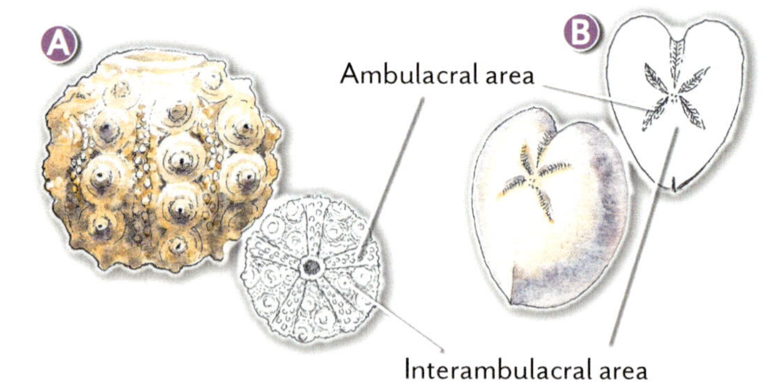

▲ Sea urchins come in regular and irregular forms: Ⓐ *Hemicidaris* is a regular echinoid from Jurassic and Cretaceous rocks. Ⓑ *Micraster* is an irregular echinoid from Upper Cretaceous rocks.

◀Ⓓ *Mesopalaeaster* is a starfish from the Ordovician of North America and Europe. Ⓔ *Palaeocoma* is a brittlestar from the Jurassic of Europe.

Starfishes (**asteroids**) and brittlestars (**ophiuroids**) are star-shaped and usually have five arms. Brittlestars tend to have longer, thinner arms than starfish.

Extinct groups include the **edrioasteroids**, which looked a bit like a starfish wrapped around a ball or disc, and the **blastoids**, which were small and bud-like in appearance.

▼ Extinct echinoderms include Ⓕ *Pentremites* (a blastoid) from the Carboniferous of North America and South America and Ⓖ *Edrioaster* (an edrioasteroid) from the Ordovician of North America and Europe.

▼ Ⓒ *Pentacrinites* is a large Triassic to Cretaceous crinoid with long arms and many branches.

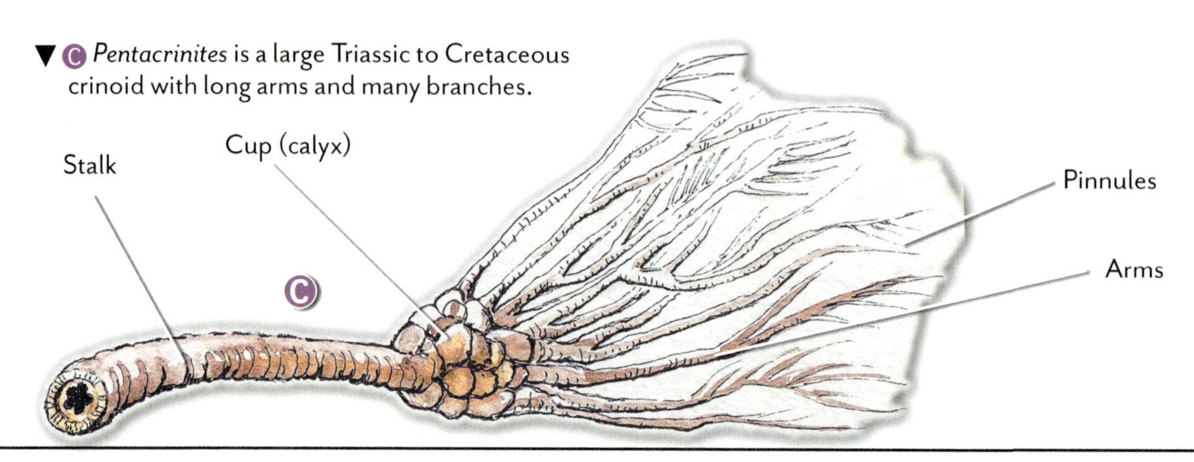

WATER VASCULAR SYSTEM OF ECHINODERMS

Echinoderms have some unique design features. One is the water vascular system, a system of water-filled canals connected to numerous tube feet. The contraction of muscles pushes the tube feet outwards through holes in the skeleton. The tube feet allow the creature to move around and gather food. Some types, such as starfish, also use their tube feet to pry open bivalve shells. If the prey animal is too large to swallow, the echinoderm can extrude its stomach to envelop it.

▼Water enters through a sieve-like structure called the madreporite before flowing into a circular canal and then along radial canals into each arm of the starfish. Both sides of each radial canal have bulb-like vessels connected to sucker-like feet. Contraction of the bulbs causes the tube feet of the starfish to extend as water is forced into them.

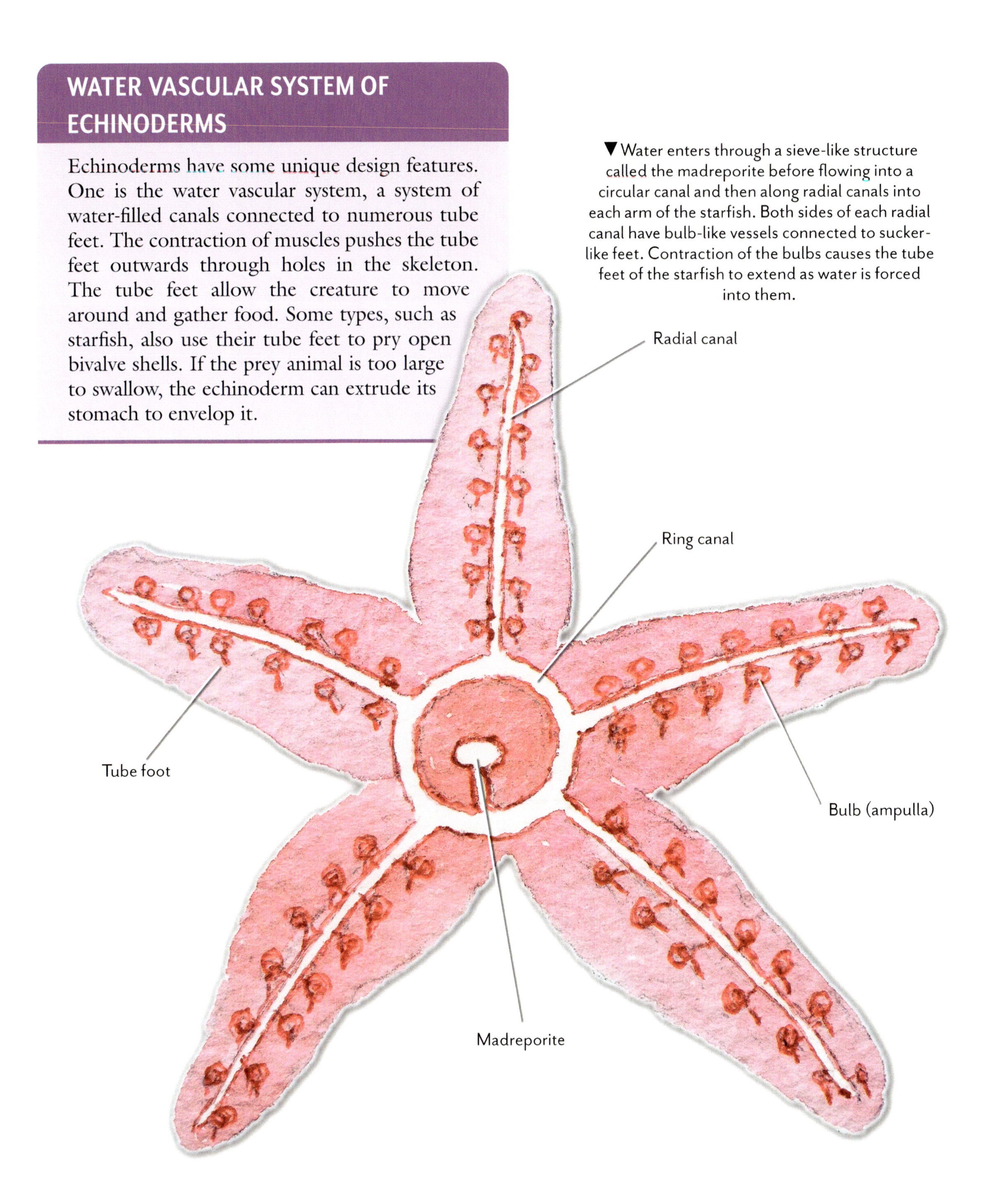

GRAPTOLITES

Graptolites are an extinct group of colonial animals. Each colony was made of branches and the individual animals (**zooids**) were housed in cups arranged along each branch. The animals probably gathered food that was floating in seawater.

There were two main graptolite groups. **Dendroids** were colonies with many branches and were probably attached to the seafloor. Their fossils are found in Cambrian to Carboniferous rocks.

Graptoloids were colonies with fewer branches and sometimes only a single branch. They were free-floating creatures and more abundant than the dendroid forms. Their fossils are found in Ordovician to Devonian rocks.

Graptolite colonies were usually quite small, not more than 4 inches (10 centimeters) long, although some types reached a couple of feet (0.5 to 1 meter).

There are no living graptolites. They did, however, have some traits similar to those of an obscure group of modern marine animals called pterobranchs. *Rhabdopleura*, a modern pterobranch, also builds branching colonies.

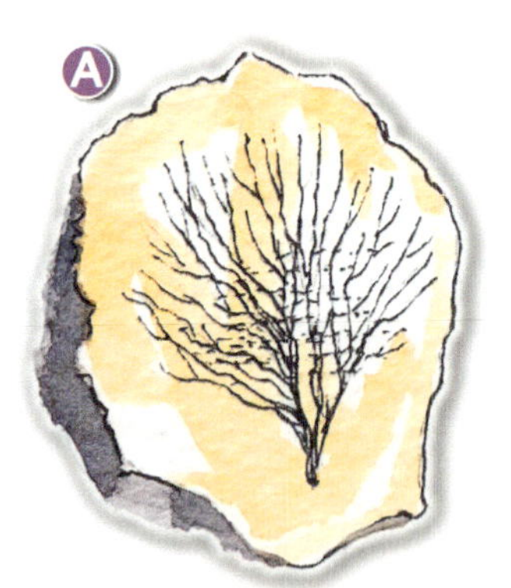

▲ Dendroid graptolites are thought to have grown like small bushes fixed to the seafloor: A *Dictyonema*, a Cambrian to Carboniferous form, was typically conical in shape. B *Acanthograptus*, a Cambrian to Silurian variety, was somewhat tree-like.

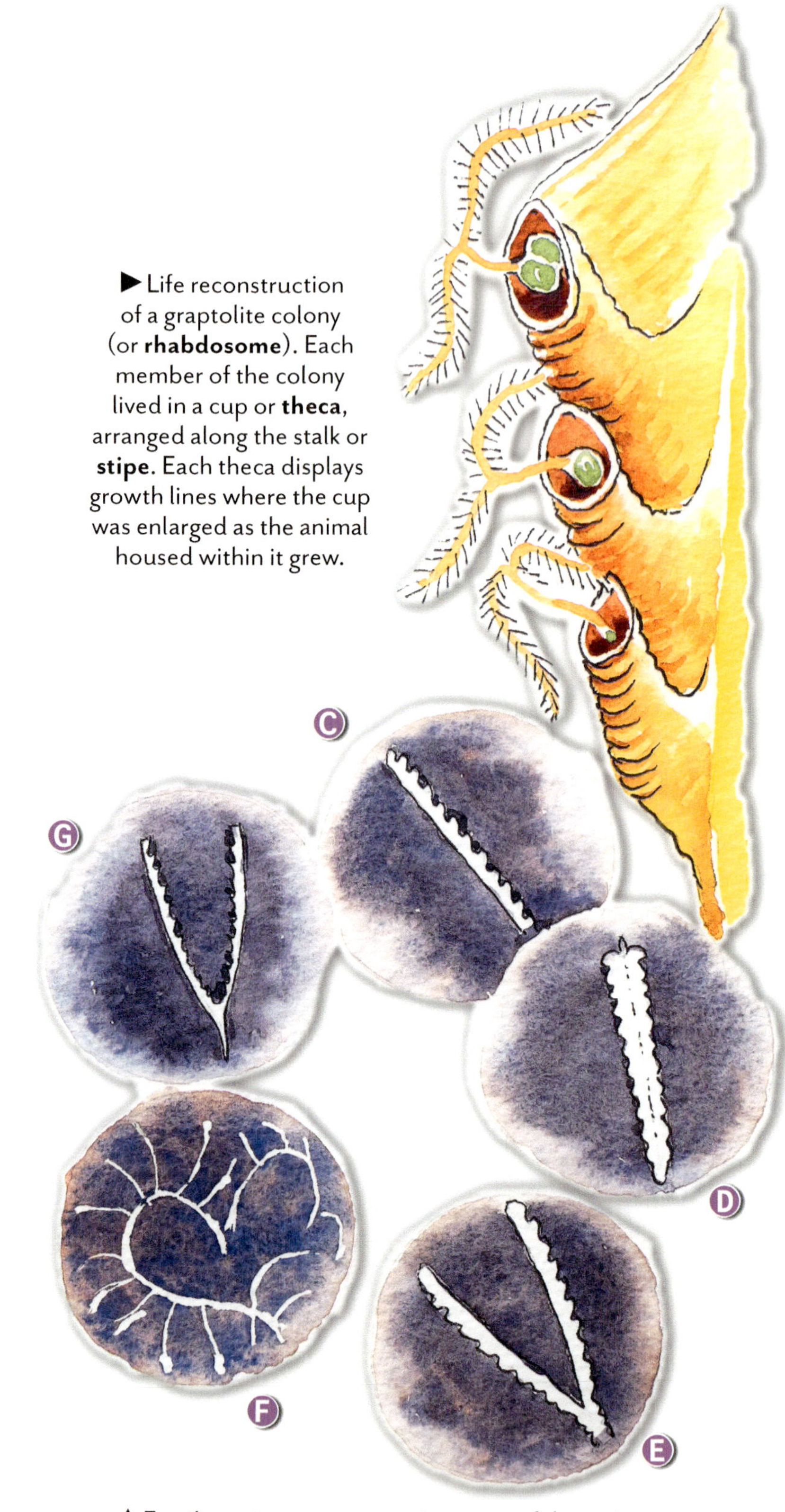

► Life reconstruction of a graptolite colony (or **rhabdosome**). Each member of the colony lived in a cup or **theca**, arranged along the stalk or **stipe**. Each theca displays growth lines where the cup was enlarged as the animal housed within it grew.

▲ Fossil specimens representing some of the major types of graptoloids: C *Monograptus*, a Silurian form in which a single row of cups is arranged along a single branch. D *Diplograptus*, an Ordovician to Silurian form in which two rows of cups are arranged along a single branch. E *Nemagraptus*, an Ordovician form with two horizontally flexed branches, each bearing a single row of cups. F *Dichograptus*, an Ordovician form with up to eight branches arranged in pairs. G *Dicranograptus*, an Ordovician form shaped a bit like a tuning fork.

VERTEBRATES I. FISHES

Vertebrates are animals with a backbone. This group includes the **fishes** and the **tetrapods** (the amphibians, reptiles, birds, and mammals).

Jawless fishes are found in rocks from the Cambrian upwards. Cambrian forms include small, streamlined animals with gills and a dorsal fin. Unfortunately, other details are hard to make out because the fossils are not sufficiently well preserved.

In Ordovician, Silurian, and Devonian rocks we find the jawless **ostracoderms**. Most seem to have been bottom-feeders, filtering particles of food from the sediments on the seafloor, although others were midwater swimmers. Some ostracoderms had bony head shields while others did not.

Fishes with jaws are found in rocks from the Silurian upwards. The **acanthodians** are especially abundant in Devonian sediments. They were mostly small, with slender bodies and large eyes. No acanthodian fossils are found after the Permian.

▼ Life reconstructions of Ⓐ *Myllokunmingia*, a probable vertebrate from the Lower Cambrian of Yunnan Province, China. It measured almost 1.2 inches (3 centimeters) from head to tail and had a sail-like fin on its back. Ⓑ *Haikouichthys* was similar to *Myllokunmingia* and is found in the same fossil beds, but it had a slightly shorter and narrower body.

▲ The jawless ostracoderms came in a variety of shapes, sizes, and life habits. Ⓒ *Hemicyclaspis* from the Silurian of North America and Europe had a bony plate enclosing the head and is thought to have been a bottom-feeder. Ⓓ *Pteraspis* from the Devonian of Great Britain had a bony head shield with a pointed snout. Ⓔ *Pharyngolepis* from the Silurian of Norway was a barrel-shaped animal that did not have a bony head shield. *Pteraspis* and *Pharyngolepis* were probably active swimmers.

▼ Ⓕ *Climatius* was a typical acanthodian from the Silurian and Devonian of Europe and North America. It had a streamlined body with many fin spines.

The armored **placoderms** are also found in Silurian to Permian rocks but are most diverse in Devonian sediments. *Coccosteus* was a small bottom-feeder but the 23-foot-long (7 meters) *Dunkleosteus* was a fearsome predator.

▲ A *Coccosteus* from the Devonian of Scotland had a bony carapace covering its head and shoulder regions. Its flattened shape suggests that it may have been a bottom-dweller.

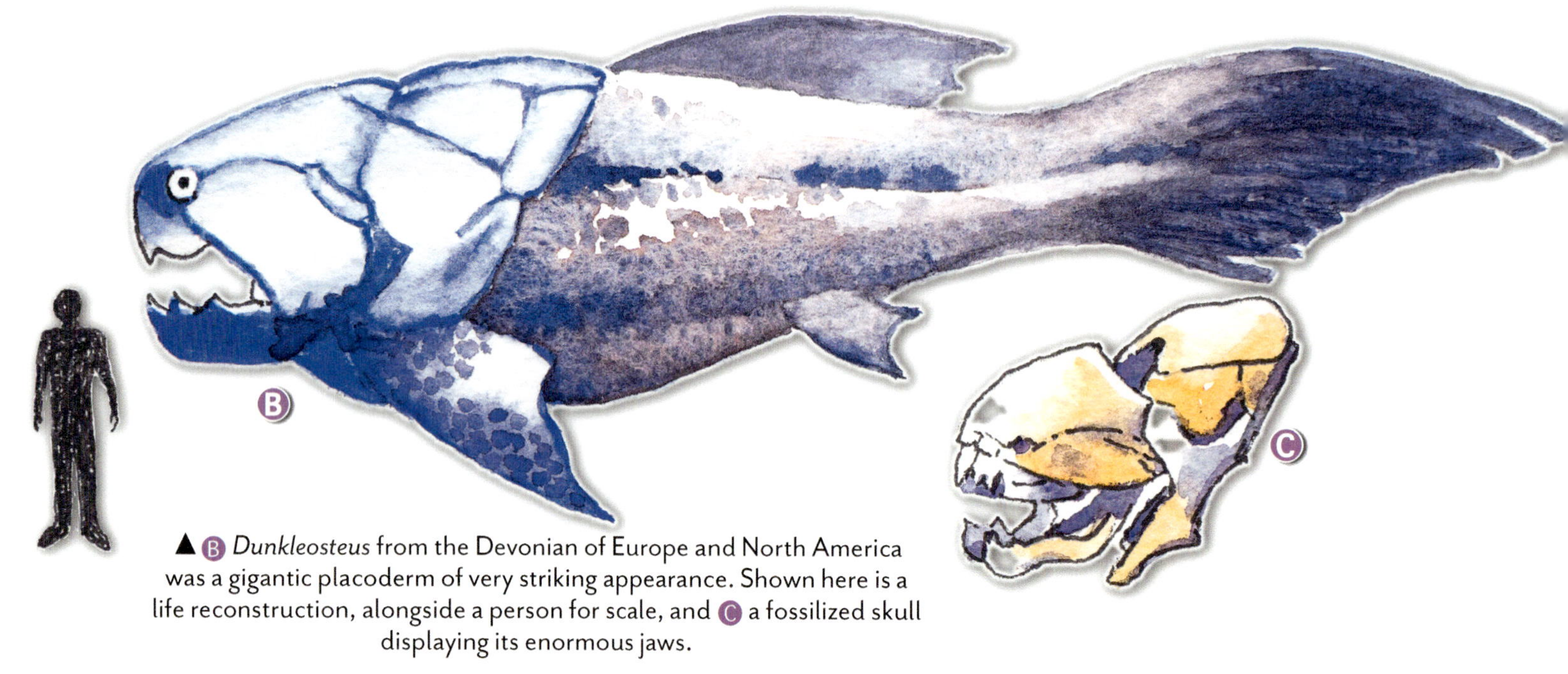

▲ B *Dunkleosteus* from the Devonian of Europe and North America was a gigantic placoderm of very striking appearance. Shown here is a life reconstruction, alongside a person for scale, and C a fossilized skull displaying its enormous jaws.

Chondrichthyans (sharks and rays) occur in rocks from the Devonian upwards. They have skeletons made of cartilage, not bone. A typical Devonian form is *Cladoselache*, which was about 6.5 feet (2 meters) long and probably a fast swimmer. Some fossilized sharks had strange spiky appendages that may have enabled males and females to clasp together during mating.

▼ D *Cladoselache* is well known from its fossilized remains in the Devonian rocks of Ohio, USA. E *Stethacanthus* is a fossilized shark found in the Devonian and Carboniferous rocks of Europe and North America. It had an unusual shoulder spine covered with small spikes or **denticles**. The purpose of this shoulder spine is not known, but it may have been for mating or display.

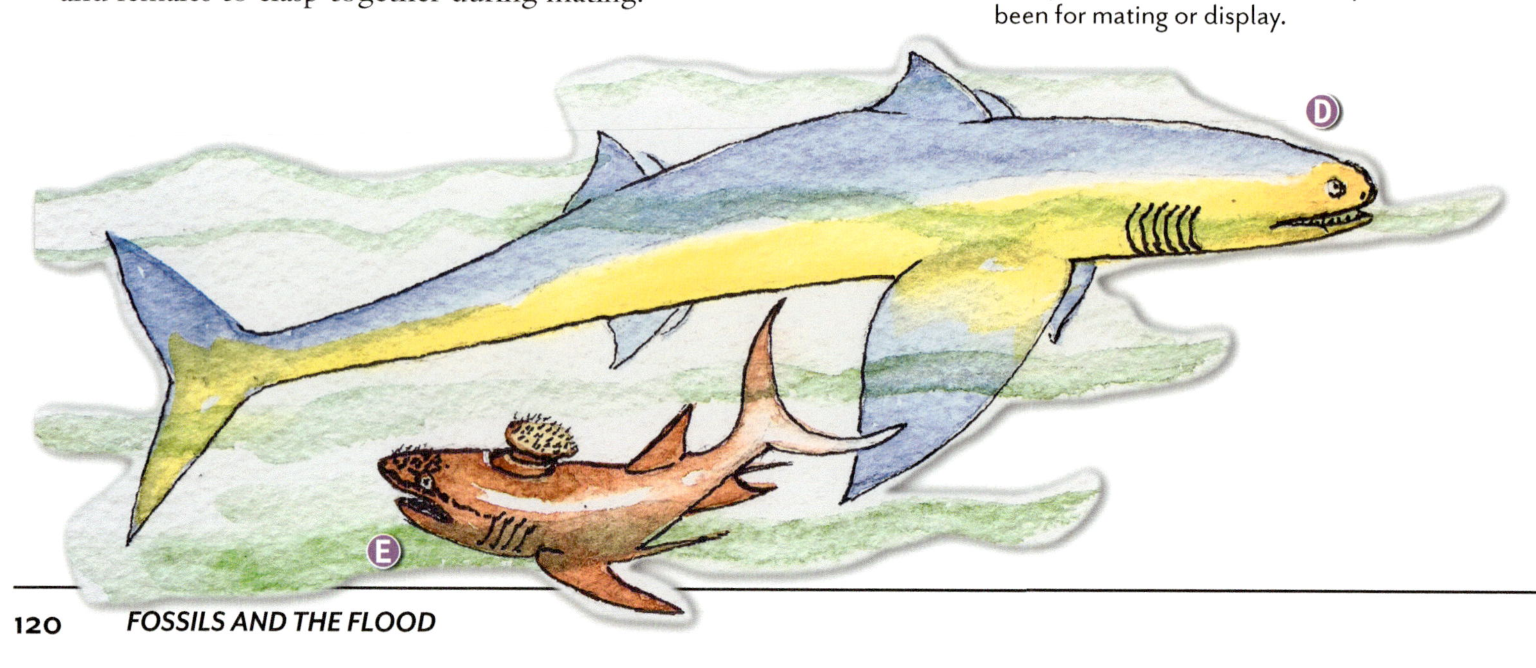

Most modern fishes are **osteichthyans** (bony fishes). Fossils representing this group range from the Devonian to the Holocene. One group of bony fishes, the **sarcopterygians** (or lobe fins), are rare today but were more abundant and diverse in the past. This group includes coelacanths and lungfishes. Another group, the **actinopterygians** (or ray fins), are mostly small fishes with noticeably forked tails and paired fins supported by thin, bony rays.

▲ F *Osteolepis* is a torpedo-shaped sarcoptyerygian with paired lobe fins. G *Cheirolepis* is a slender-bodied actinopterygian with triangular dorsal and anal fins. Both are from the Middle Devonian of Scotland.

THE CONODONT MYSTERY

For many years scientists were puzzled by the discovery of small, tooth-like fossils in Cambrian to Triassic marine rocks. Some thought that these conodont elements were bits of bivalves, sponges, or worms. Others thought they had belonged to a vertebrate. The mystery was solved when the first complete animal was found (*Clydagnathus* from the Carboniferous of Edinburgh, Scotland). The conodont elements turned out to be the teeth of extinct, eel-like fishes, now known as conodontophores (meaning *conodont bearers*).

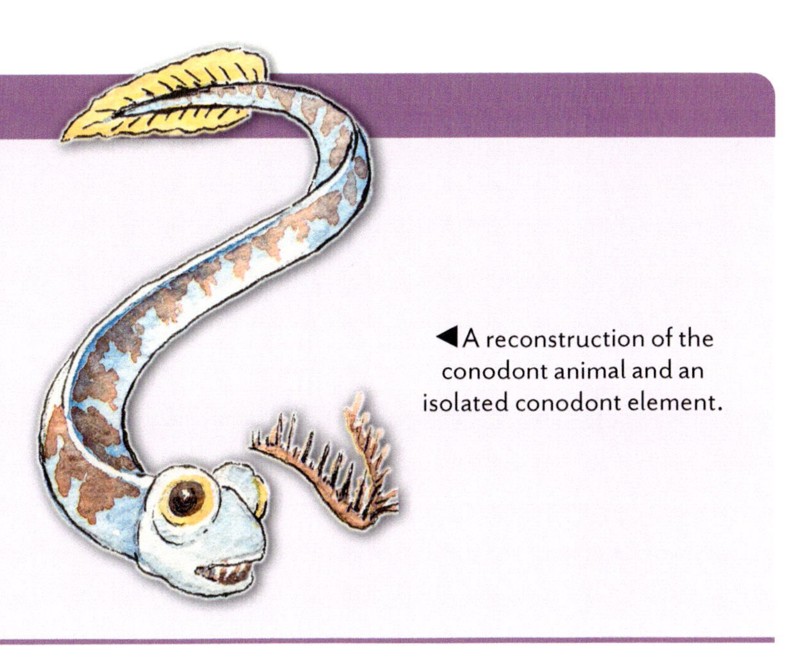

◀A reconstruction of the conodont animal and an isolated conodont element.

VERTEBRATES II. AMPHIBIANS AND REPTILES

Tetrapods (from the Greek *tetra* meaning *four* and *pod* meaning *foot*) are vertebrates with legs and digits, although some groups (such as snakes and caecilians) are legless. Included among the tetrapods are the amphibians, reptiles, birds, and mammals. Here we focus upon the **amphibians** and **reptiles**.

In Upper Devonian rocks we find semi-aquatic tetrapods that were up to 3.3 feet (1 meter) or so in length and probably preyed on fishes. They had fish-like tails, internal gills, and paddle-like limbs with more than five digits.

▼*Acanthostega* and *Ichthyostega* are semi-aquatic tetrapods from the Upper Devonian of Greenland. A *Acanthostega* was just over 1.64 feet (0.5 meters) long and had eight digits on each forelimb. B *Ichthyostega* was about 5 feet (1.5 meters) long and had seven digits on each hind limb.

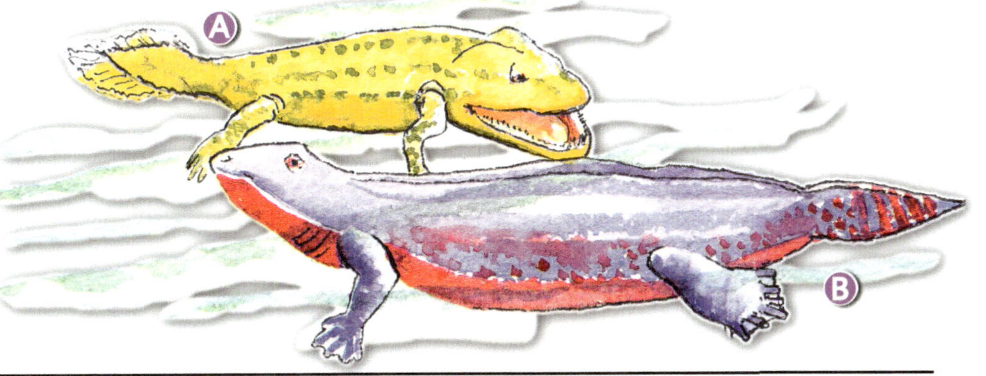

Carboniferous, Permian, and Triassic rocks yield a diverse array of amphibians, ranging from small, lizard-like forms to larger aquatic and semi-aquatic forms. Some of these fossil groups range into the Jurassic and Cretaceous, but most do not extend beyond the Triassic.

The **lissamphibia**, the group that includes modern amphibians such as frogs and toads, newts and salamanders, and caecilians, has a fossil record spanning the Triassic to the Holocene.

▶ C Other temnospondyls were much larger, such as *Prionosuchus* from the Permian of Brazil. This animal is estimated to have reached 30 feet (9 meters) long and had a pointed snout with sharp teeth. D Another group, the lepospondyls, included *Diplocaulus*. It was 3.3 feet (1 meter) long and had a bizarre boomerang-shaped head. Its fossil remains are found in the Permian rocks of North America and Africa.

▼ Many types of extinct amphibians are known from the fossil record. The temnospondyls included animals such as A *Cochleosaurus* (about 5 feet or 1.5 meters long) and B *Dendrerpeton* (about 3.3 feet or 1 meter long), both from the Carboniferous of Nova Scotia, Canada.

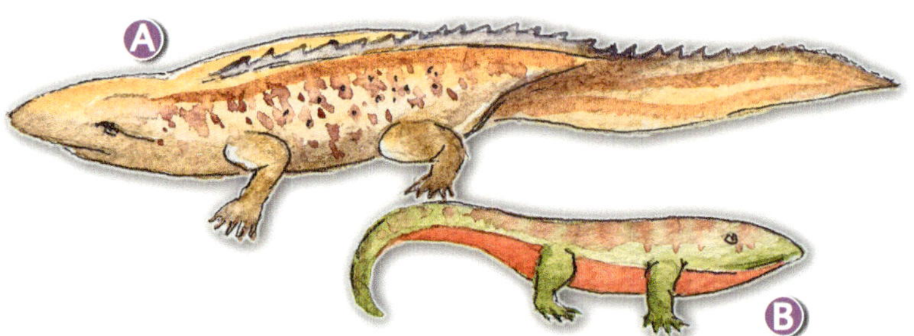

▼ E *Gephyrostegus* was an anthracosaur from the Carboniferous of Europe. It was about 10 inches (25 centimeters) long and probably fed on insects.

▶ F *Triadobatrachus* was an extinct frog-like animal from the Triassic of Madagascar. Its skull was similar to that of modern frogs but it had more vertebrae in its backbone, six of which formed a short tail. It probably did not jump like modern frogs.

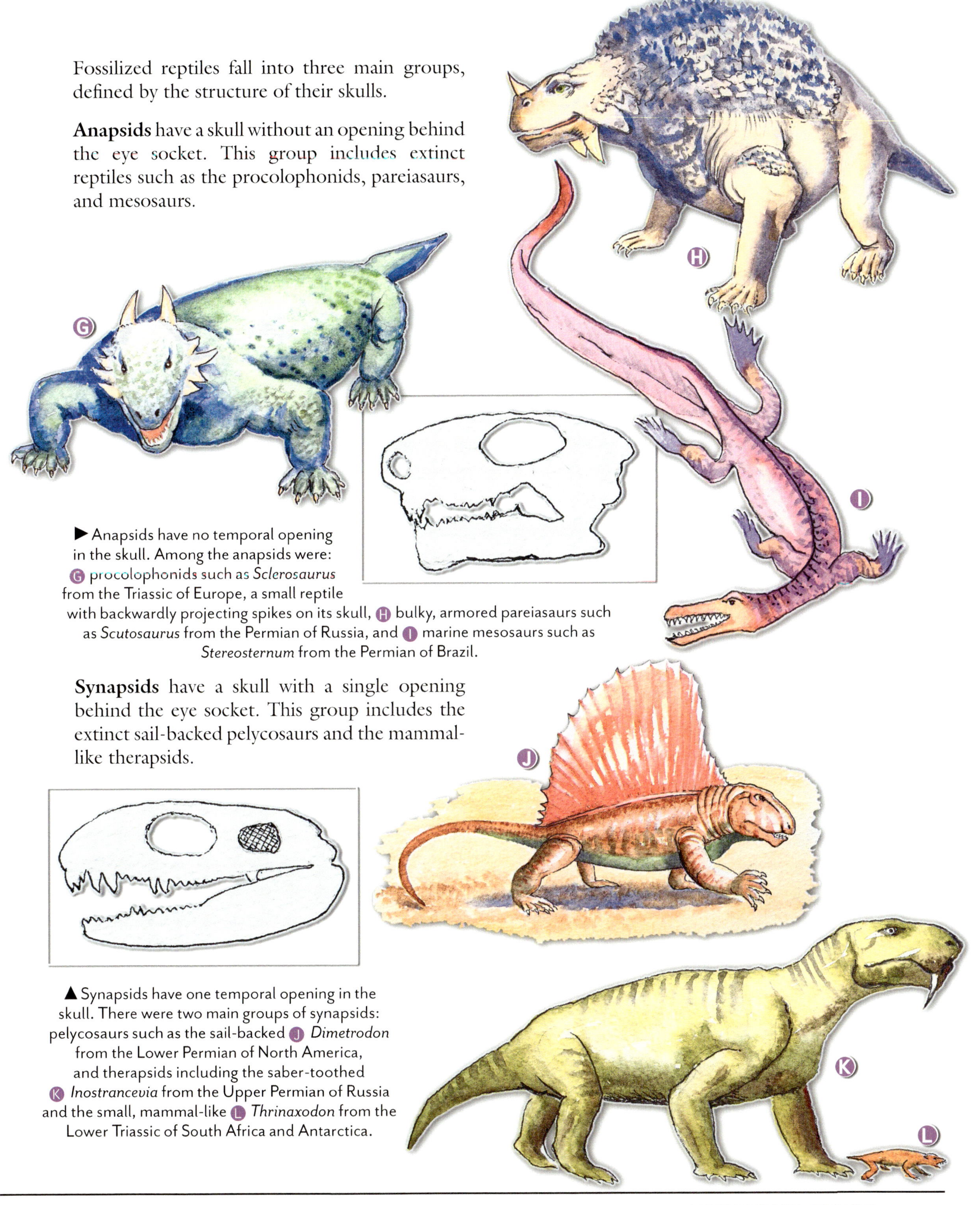

Fossilized reptiles fall into three main groups, defined by the structure of their skulls.

Anapsids have a skull without an opening behind the eye socket. This group includes extinct reptiles such as the procolophonids, pareiasaurs, and mesosaurs.

► Anapsids have no temporal opening in the skull. Among the anapsids were: G procolophonids such as *Sclerosaurus* from the Triassic of Europe, a small reptile with backwardly projecting spikes on its skull, H bulky, armored pareiasaurs such as *Scutosaurus* from the Permian of Russia, and I marine mesosaurs such as *Stereosternum* from the Permian of Brazil.

Synapsids have a skull with a single opening behind the eye socket. This group includes the extinct sail-backed pelycosaurs and the mammal-like therapsids.

▲ Synapsids have one temporal opening in the skull. There were two main groups of synapsids: pelycosaurs such as the sail-backed J *Dimetrodon* from the Lower Permian of North America, and therapsids including the saber-toothed K *Inostrancevia* from the Upper Permian of Russia and the small, mammal-like L *Thrinaxodon* from the Lower Triassic of South Africa and Antarctica.

The **diapsids** have a skull with two openings behind the eye socket. This group includes lepidosaurs (lizards, snakes, and tuataras) and archosaurs (crocodiles, alligators, pterosaurs, and dinosaurs). Turtles are also classified as diapsids, although they do not have the typical diapsid skull.

The synapsids and anapsids are most abundant and diverse in Carboniferous to Permian sediments and the diapsids are most common in Triassic, Jurassic, and Cretaceous sediments.

▼Diapsids have two temporal openings in the skull. Most spectacular among the diapsids were the dinosaurs. Representatives of the main dinosaur groups are shown here:

- A *Plateosaurus* (prosauropod).
- B *Staurikosaurus* (herrerasaur).
- C *Pentaceratops* (ceratopsian).
- D *Lambeosaurus* (hadrosaur).
- E *Sauropelta* (ankylosaur).
- F *Kentrosaurus* (stegosaur).
- G *Carcharodontosaurus* (theropod).
- H *Alaskacephale* (pachycephalosaur).
- I *Brachiosaurus* (sauropod).

VERTEBRATES III. BIRDS AND MAMMALS

The other tetrapods are the birds and mammals.

Birds are rarely found as fossils because their bones are delicate and do not preserve very well. Nevertheless, representatives of the main groups are sometimes found in the fossil record.

Extinct, toothed birds are known from Jurassic and Cretaceous rocks. **Enantiornithines** (meaning *opposite birds*) were typically about the size of sparrows and starlings, although some were larger. Other toothed birds were the fish-eating *Hesperornis* and *Ichthyornis*, both from the Upper Cretaceous of North America.

Palaeognaths are flightless birds, represented today by the emus and cassowaries of Australia, the ostriches of Africa, the rheas of South America, and the kiwis of New Zealand. Extinct forms include *Lithornis* from the Paleocene of Wyoming, USA, and *Palaeotis* from the Eocene of Germany. There were also giant varieties such as the moas of New Zealand and the elephant birds of Madagascar.

▲ A *Sinornis* is known from Lower Cretaceous rocks in China. It was about the size of a sparrow. B *Enantiornis* resembled a vulture in size and habits. It is known from Upper Cretaceous fossil beds in Argentina.

▼ C *Hesperornis* was a large, flightless, diving bird, about 5.9 feet (1.8 meters) long. It probably swam using its powerful back legs. D *Ichthyornis* was a smaller bird and probably had tern- or gull-like habits.

▼ E *Lithornis* is from the Paleocene and Eocene of North America and Europe. It was about the size of a chicken. F *Palaeotis* is from the Eocene of Europe. It was about the size of a crane.

◀ G New Zealand was once home to giant flightless birds called moas, the largest of which may have reached 12 feet (3.6 meters) in height. They became extinct in the fifteenth century. H Large flightless birds called elephant birds also lived on the island of Madagascar. They became extinct in the seventeenth or eighteenth century.

Most living birds are classified as **neognaths**. This is a very diverse group including ducks and waterfowl, diving birds and waders, raptors and vultures, songbirds, hummingbirds, and woodpeckers. There is uncertainty about how far back in the fossil record these groups can be traced, but some may be represented in the Upper Cretaceous.

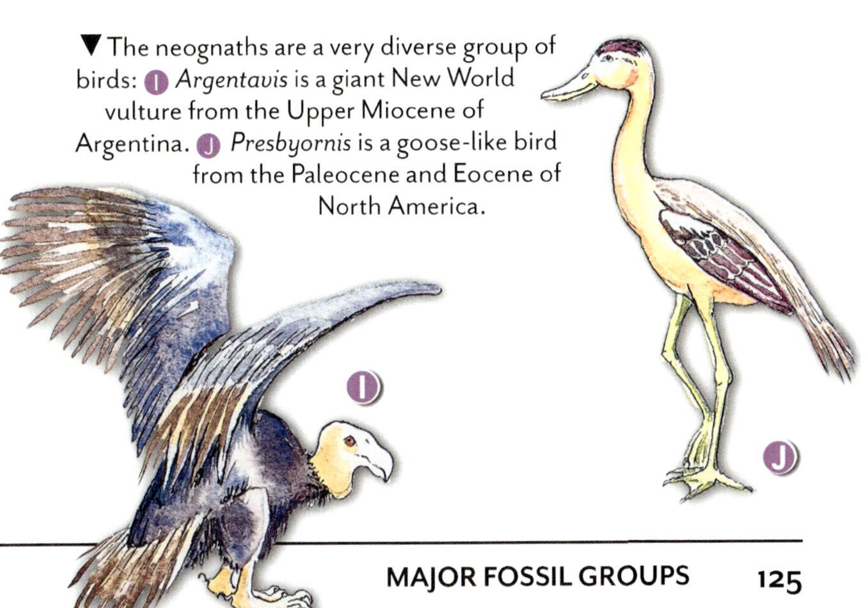

▼ The neognaths are a very diverse group of birds: I *Argentavis* is a giant New World vulture from the Upper Miocene of Argentina. J *Presbyornis* is a goose-like bird from the Paleocene and Eocene of North America.

Mammals are distinguished from modern reptiles by the possession of hair and mammary glands. Fragmentary fossil remains have been discovered in Upper Triassic rocks, but better known are the shrew-like morganucodonts and megazostrodonts from the Lower Jurassic, and the rodent-like multituberculates that range from the Upper Jurassic to Lower Oligocene.

Living mammals are classified in three groups: the **monotremes**, **marsupials**, and **placentals**. We know monotremes today as the duck-billed platypus of Australia and the echidnas of Australia and New Guinea. In the mid-1980s, monotreme-like jaw fragments were described from Australia. They extended the fossil record of this group back to the Lower Cretaceous.

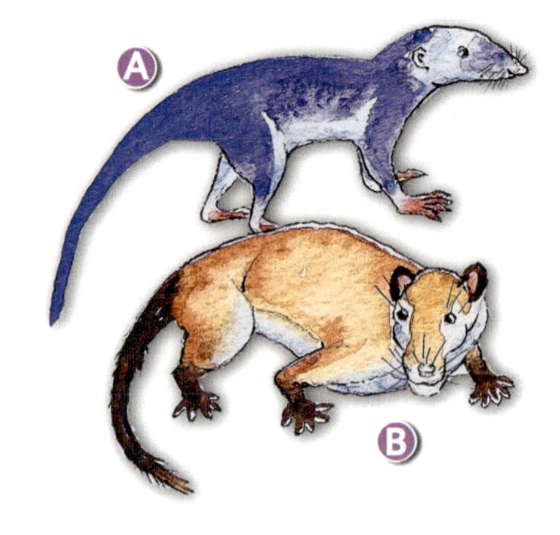

◀ A *Morganucodon* and B *Megazostrodon* are small, shrew-like animals known from Upper Triassic and Lower Jurassic rocks.

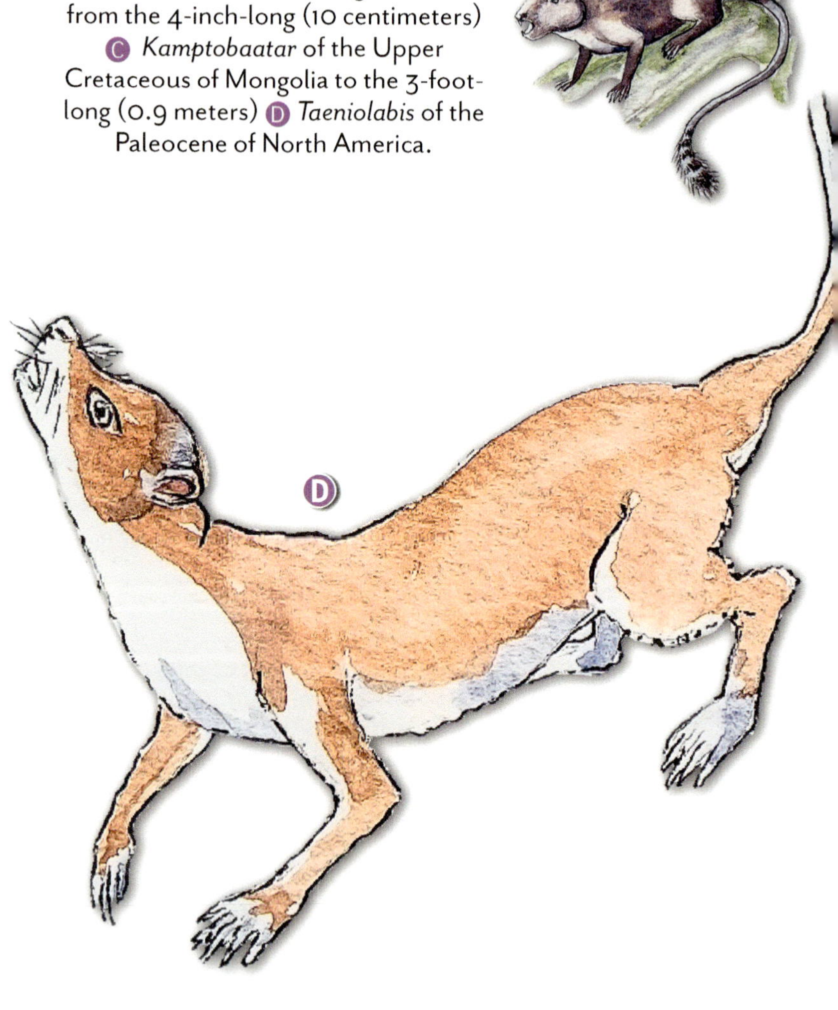

◀ Multituberculates ranged in size from the 4-inch-long (10 centimeters) C *Kamptobaatar* of the Upper Cretaceous of Mongolia to the 3-foot-long (0.9 meters) D *Taeniolabis* of the Paleocene of North America.

▼ E *Obdurodon dicksoni* is a Miocene platypus reconstructed from a complete skull preserved in the famous Riversleigh fossil beds of northern Australia. It is estimated to have been about 2 feet (0.6 meters) long from head to tail.

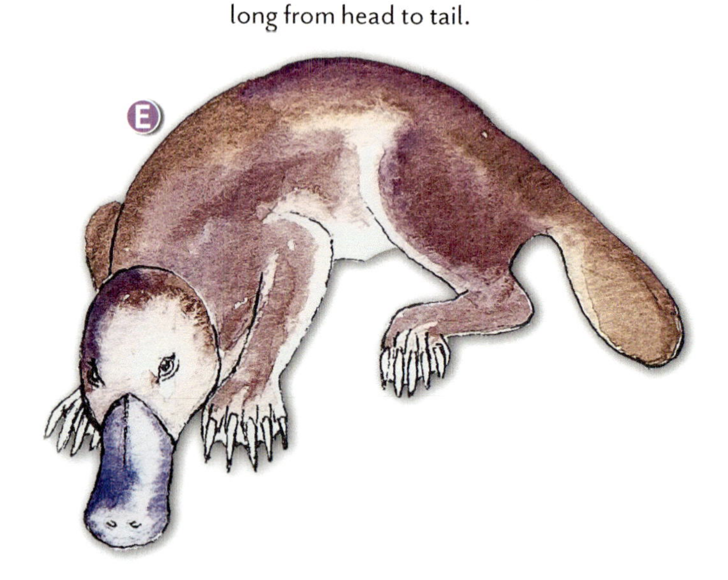

Marsupials occur only in Australasia and South America today, but their fossils have also been found in Africa, Asia, Antarctica, and North America. They have a fossil record beginning in the Middle Cretaceous, although most of the Cretaceous fossils are jaw fragments and teeth.

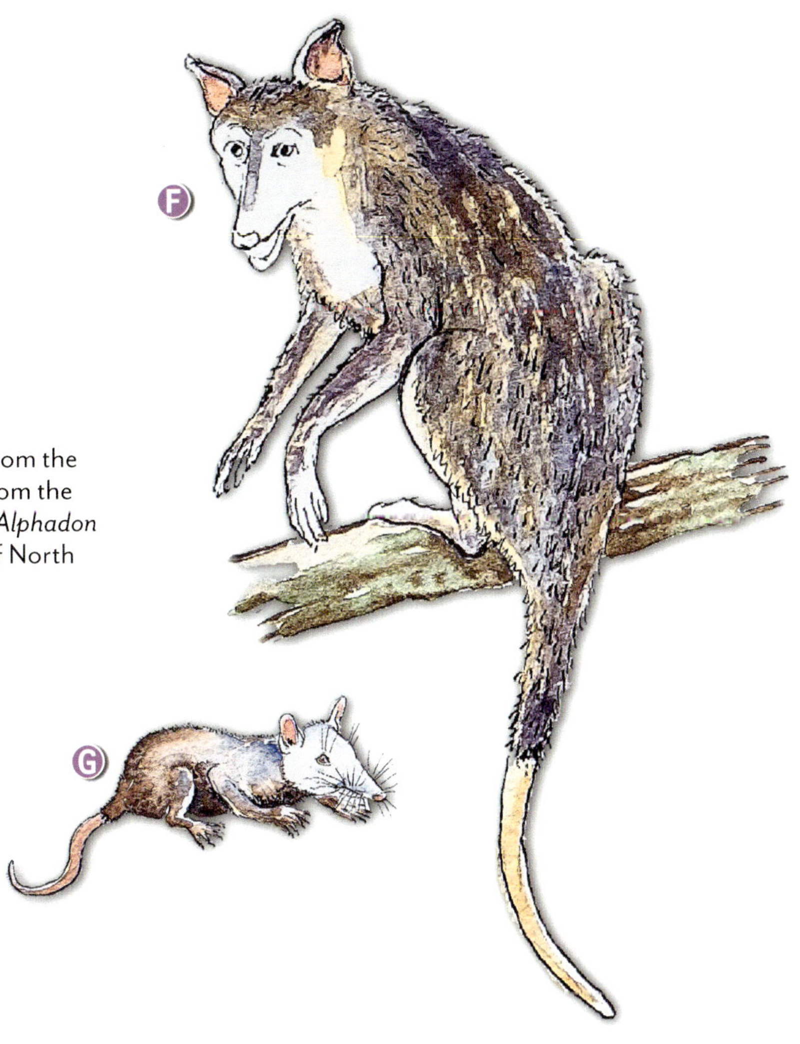

▶ Opossum-like marsupials from the Cretaceous: F *Sinodelphys* from the Lower Cretaceous of China. G *Alphadon* from the Upper Cretaceous of North America.

Placentals are the dominant mammals today. The most diverse groups are the bats and the rodents. Other modern groups include carnivores such as cats, dogs, and bears and herbivores such as horses, camels, and hippopotamuses. There are also aquatic and semi-aquatic placentals such as whales, seals, and sea lions. Fossil placentals extend back to the Upper Cretaceous and many now-extinct groups are recognized.

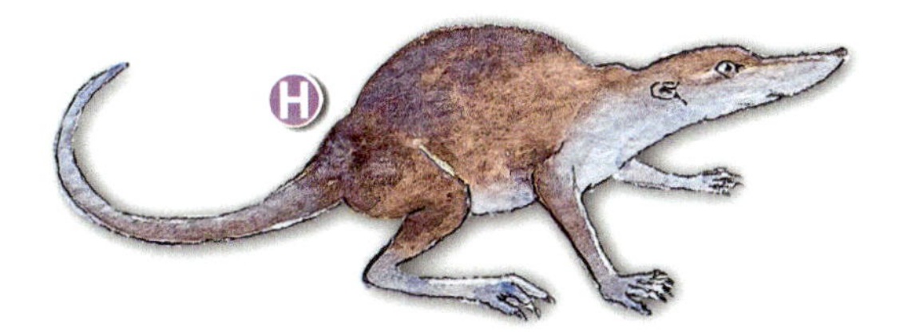

▲ One of the first placentals in the fossil record is the Upper Cretaceous H *Zalambdalestes*.

ODD-TOED AND EVEN-TOED UNGULATES

One major group of placentals are the **ungulates**, or hoofed mammals. Ungulates are divided into **perissodactyls** and **artiodactyls**. Perissodactyls (such as horses, tapirs, and rhinoceroses) have an odd number of toes. Artiodactyls (such as deer, camels, cattle, and pigs) have an even number of toes. Many extinct types belonging to these groups are also known.

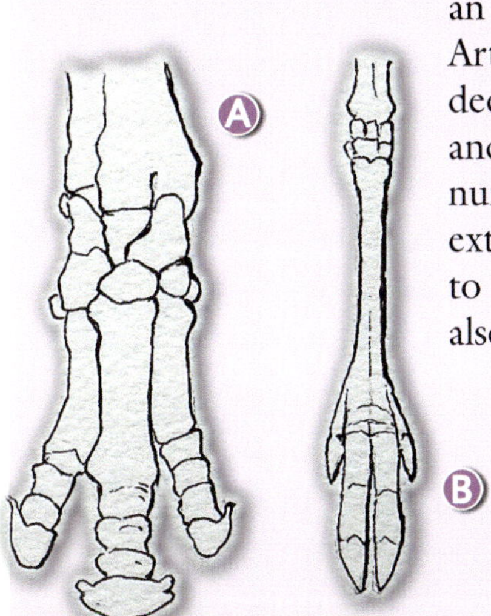

◀ A Rhinoceroses are perissodactyls and have three toes. B Deer are artiodactyls and have two toes.

A PERSONAL REFLECTION: GLORIFYING GOD THROUGH SCIENTIFIC DISCOVERY

We have now concluded our journey of discovery through the fossil record. We have seen how the historical accounts in the Bible, together with the insights provided by fossils and other types of scientific evidence, allow us to reconstruct the early history of the earth. Perhaps this book has inspired you to want to learn more about fossils and what they can tell us. It was reading books like this one that got me interested in fossils, too.

In 1987, I enrolled as an undergraduate student on a combined science program in biology and geology. I majored in geology and graduated in 1990. Later in life, I undertook a masters program with a paleobiology emphasis—studying the large-scale patterns and processes in the history of life that can be deduced from the fossil record. These programs gave me an invaluable grounding in both the subject matter of my discipline and the scientific method. I am grateful to my teachers for all they passed on to me even though, to my knowledge, none of them shared my creationist convictions.

Since 2002 I have been a researcher with Biblical Creation Trust, a small creationist organization based in the United Kingdom. My work has given me the opportunity to get to know many research scientists who accept the Bible's account of earth history as I do. I have been privileged to work with some of them, including my good friends Dr. John Whitmore (senior professor of geology at Cedarville University in Ohio, USA) and Ray Strom (president of Calgary Rock and Materials Services Inc. in Canada). The three of us have spent many hours in the field together studying the Coconino Sandstone and other rock units in the southwestern United States. We have shared many memorable experiences—hiking in the Grand Canyon and other wilderness areas, making scientific measurements and collecting rock samples, and all the while doing our best to avoid rattlesnakes and other hazards! Our last field season together on that project was in 2011, but the work has gone on since, not least in the capable hands of some of Dr. Whitmore's fine students.

Some people may consider it strange that Christians would want to devote their lives to the study of the natural world. In fact, Christians have the best reason of all—we know the Creator who made the world and everything in it! The Bible tells us that the invisible attributes of God are clearly seen through the things he has made (Romans 1:20). Just as we are able to know an artist better through studying his paintings or an author by reading his books, so we are able to grow in our knowledge of God by studying his creation. And as we get to know him better, we are able to glorify him all the more, with greater love and devotion.

There are other reasons for Christians to practice science too. The study of God's world helps us to fulfill our God-given mandate to care for the creation and to minister to others. At the beginning of Creation, God gave human beings the authority to rule over the earth and all its creatures (Genesis 1:28). But in order to be wise stewards of the creation we need to understand how it works. Science is a powerful tool for gaining that understanding. Moreover, we can apply our scientific knowledge to benefit others, for example in treating disease, developing life-enhancing technologies, and addressing environmental problems.

Of course, the scientific study of God's world is not easy—few truly worthwhile things are! Science requires patience, determination, and hard work—but it can also be a lot of fun! One of my favorite Bible verses is Psalm 111:2: "The works of the LORD are great, sought out of all them that have pleasure therein." The psalmist reminds us that we are to delight in the works of God—to relish and enjoy them! There are many ways to do this—but one excellent way is through scientific enquiry.

As we study God's creation, one of the most amazing things we learn is that we will never run out of new things to discover. No matter how deep we dig into God's creation there are always unfathomable depths still to explore. There is so much we do not know, and for every question our research answers, many more are raised. In this book I have sought to draw upon the best creationist scholarship available in reconstructing the earth's early history. But the story is far from complete. Even granting that the basic outline is correct, the details are often very uncertain. We still have a huge amount to learn about such things as the world before the Flood, the processes that took place during the Flood, and the recovery of the world after the Flood.

Perhaps this is where you come in. Are you a talented and motivated student with an aptitude for science and a willingness to work hard? Have you considered studying a relevant subject at college? We need more scholars in biology, geology, astronomy, and other areas. Perhaps you could even specialize in a field like paleontology or radiometric dating. Getting the very best training in these disciplines may well mean studying in an evolutionary department in a secular university—and that's not easy for a creationist student. But these disciplines are often the most valuable as we seek to develop, test, and refine our creationist understanding of earth history.

Of course, much of the life of a scientist is spent working away at small pieces of the overall puzzle, and moments of breakthrough—when someone comes up with something truly revolutionary—occur only rarely. But whether the contribution that we make is large or small, the scientific life is richly rewarding. And even if the scientific life is not for you, perhaps you can encourage a fellow Christian for whom it is. Being a creationist can be hard, especially in a secular university setting, and it will make all the difference to someone in this situation to know that they have the prayerful support of others around them.

Whatever your calling, my prayer is that this book will have helped you to look afresh at the wonders in the world around you and imparted something of the excitement and joy of scientific discovery as a means of bringing glory to our Creator God. May he richly bless you as you serve him.

“And God said, This is the token of the covenant which I make between me and you and every living creature that is with you, for perpetual generations: I do set my bow in the cloud, and it shall be for a token of a covenant between me and the earth.

And it shall come to pass, when I bring a cloud over the earth, that the bow shall be seen in the cloud: And I will remember my covenant, which is between me and you and every living creature of all flesh; and the waters shall no more become a flood to destroy all flesh."

Genesis 9:12-15

RECOMMENDED RESOURCES

For parents and educators

Visit our web page at www.fossilsandtheflood.net to find supportive material in the form of notes and references for each chapter of this book. These notes and references are intended to help you understand more fully the biblical and scientific evidence supporting the claims made in each chapter. They also give guidance for further exploration of the published literature. Please note that the references are only illustrative and not intended to be exhaustive. A more select bibliography can be found below.

Books

Austin, S.A. (editor). 1994. *Grand Canyon: Monument to Catastrophe.* Institute for Creation Research, Santee, California. ISBN 0-932766-33-1.

Brand, L. 2005. *Beginnings: Are Science and Scripture Partners in the Search for Origins?* Pacific Press Publishing Association, Nampa, Idaho. ISBN 0-8163-2144-2.

Brand, L. and Chadwick, A. 2016. *Faith, Reason, and Earth History: A Paradigm of Earth and Biological Origins by Intelligent Design.* Third edition. Andrews University Press, Berrien Springs, Michigan. ISBN 978-1-940980-11-9.

Coffin, H., Brown, R.H., and Gibson, L.J. 2005. *Origin by Design.* Revised Edition. Review and Herald Publishing Association, Hagerstown, Maryland. ISBN 0-8280-1776-X.

DeYoung, D. 2005. *Thousands … Not Billions: Challenging an Icon of Evolution, Questioning the Age of the Earth.* Master Books, Green Forest, Arkansas. ISBN 0-89051-441-0.

Garner, P. 2009. *The New Creationism: Building Scientific Theories on a Biblical Foundation.* Evangelical Press, Darlington. ISBN 0-85234-692-1.

Oard, M.J. 1990. *An Ice Age Caused by the Genesis Flood.* Institute for Creation Research, El Cajon, California. ISBN 0-932766-20-5.

Snelling, A.A. 2009. *Earth's Catastrophic Past: Geology, Creation and the Flood*: 2 volumes. Institute for Creation Research, Dallas, Texas. ISBN 0-890518-74-8.

Wise, K.P. 2002. *Faith, Form, and Time: What the Bible Teaches and Science Confirms about Creation and the Age of the Universe.* Broadman and Holman Publishers, Nashville, Tennessee. ISBN 0-8054-2462-8.

Wise, K.P. and Richardson, S.A. 2004. *Something from Nothing: Understanding What you Believe about Creation and Why.* Broadman and Holman Publishers, Nashville, Tennessee. ISBN 0-8054-2779-1.

Wood, T.C. 2018. *The Quest: Exploring Creation's Hardest Problems.* Compass Classroom, Nashville, Tennessee. ISBN 978-0-9990409-4-2.

Wood, T.C. and Murray, M.J. 2003. *Understanding the Pattern of Life: Origins and Organization of the Species.* Broadman and Holman Publishers, Nashville, Tennessee. ISBN 0-8054-2714-7.

DVDs

Is Genesis History? Featuring Del Tackett, Kevin Anderson, Steven Austin, Steven Boyd, Robert Carter, Arthur Chadwick, Danny Faulkner, George Grant, Paul Nelson, Douglas Petrovich, Marcus Ross, Andrew Snelling, Kurt Wise, and Todd Wood. 100 minutes, Compass Cinema, 2017.

Beyond Is Genesis History? Volume 1: Rocks & Fossils. Featuring Del Tackett, Steven Austin, Marcus Ross, Kurt Wise, Larry Vardiman, Arthur Chadwick, and Andrew Snelling. Compass Cinema, 2017.

Beyond Is Genesis History? Volume 2: Life & Design. Featuring Del Tackett, Todd Wood, Kurt Wise, Stuart Burgess, Robert Carter, Kevin Anderson, and Paul Nelson. Compass Cinema, 2018.

Beyond Is Genesis History? Volume 3: Bible & Stars. Featuring Del Tackett, Steven Boyd, George Grant, Danny Faulkner, Douglas Petrovich, and Douglas Kelly. Compass Cinema, 2018.

Set in Stone. Featuring Paul Garner, John Whitmore, and Andrew Snelling. 58 minutes, Truth in Science, 2012.

Periodicals

Acts & Facts (icr.org/aaf)

Answers Magazine (answersingenesis.org/answers/magazine)

Answers Research Journal (answersresearchjournal.org)

Creation Magazine (creation.com/creation-magazine-articles)

Creation Research Society Quarterly (creationresearch.org/crsq-journal)

e-Origins (biblicalcreationtrust.org/resources-e-origins.html)

Journal of Creation (creation.com/journal-of-creation-articles)

Journal of Creation Theology and Science Series B: Life Sciences (coresci.org/jcts/index.php/jctsb)

Journal of Creation Theology and Science Series C: Earth Sciences (coresci.org/jcts/index.php/jctsc)

Origins (Geoscience Research Institute) (grisda.org/origins-1)

Websites

Answers in Genesis (answersingenesis.org)

Biblical Creation Trust (biblicalcreationtrust.org)

Core Academy of Science (coresci.org)

Creation Biology Society (creationbiology.org)

Creation Ministries International (creation.com)

Creation Research (creationresearch.net)

Creation Research Society (creationresearch.org)

Creation Theology Society (creationtheologysociety.org)

Geoscience Research Institute (grisda.org)

Human Genesis (humangenesis.org)

Institute for Creation Research (icr.org)

International Conference on Creationism (internationalconferenceoncreationism.com)

Is Genesis History? (isgenesishistory.com)

New Creation Blog (newcreation.blog)

Podcast

Let's Talk Creation (coresci.org/podcast). A fortnightly podcast featuring Paul Garner and Todd Wood, with episodes released on YouTube and a range of streaming platforms.

GLOSSARY

acanthodians. An extinct group of jawed fishes with streamlined bodies and prominent spines.

actinopterygians. A group of bony fishes in which the fins are supported by thin, bony rays. Commonly known as ray fins.

Agnostida. A group of small trilobites, usually lacking eyes, and with only two or three segments in the thorax.

algae. A general term for plant-like, single-celled microorganisms that harness sunlight to produce nutrients.

ammonites. A group of ammonoids with complex sutures showing frilly saddles and lobes.

ammonoids. An extinct group of cephalopod mollusks. Most had tightly coiled shells, but some were uncoiled or partially coiled. Includes goniatites, ceratites, and ammonites.

amphibians. A group of tetrapods distinguished by having a gill-breathing larval stage, usually followed by a lung-breathing adult stage.

anapsids. A group of tetrapods with no openings in the skull behind the eye. The extinct procolophonids, pareiasaurs, and mesosaurs are types of anapsids.

angiosperms. A major group of plants with seeds enclosed in fruits. Commonly known as flowering plants.

animals. A large and diverse group of organisms that feed on organic matter and typically have specialized sense organs that allow them to rapidly respond to stimuli.

anomalodesmatans. A group of burrowing or boring bivalves, often with somewhat elongated shells.

arthropods. A major group of invertebrates with hard external skeletons made of chitin, segmented bodies, and jointed limbs. Insects, spiders, crabs, and trilobites are types of arthropods.

articulates. A group of brachiopods, each with a toothed hinge.

artiodactyls. A group of hoofed mammals, each with an even number of toes. Pigs, deer, cattle, and camels are types of artiodactyls.

Asaphida. A group of trilobites, each with a large, smooth cephalon and pygidium, and typically five to twelve segments in the thorax.

asteroids. A group of echinoderms with five distinct arms. Commonly known as starfish.

Atdabanian. A subdivision of Cambrian rocks, distinguished by the first appearance of trilobite fossils.

axis. The central lobe of a trilobite.

Babel. The city and tower built on the plain of Shinar after the Flood. God's intervention at Babel led to the confusion of languages and the dispersion of humanity.

bacteria. A group of single-celled microorganisms that have cell walls but lack nuclei.

baramins. A word meaning *created kinds*; basic types of organisms created separately from other kinds.

baraminology. An explicitly creationist method of classifying and categorizing organisms proposed by Kurt Wise in 1990.

belemnoids. An extinct group of cephalopod mollusks with squid-like soft parts. Their fossil remains consist mostly of the bullet-shaped internal shell called the rostrum.

Bennettitales. An extinct group of cycad-like gymnosperms.

benthic. Inhabiting the bottom of a sea or lake, whether burrowing into, crawling on, or fixed to the bottom.

biology. The scientific study of living organisms.

biome. A major ecological community of plants and animals extending over a large region.

birds. A group of tetrapods that possess feathers, wings, and a beak, and lay hard-shelled eggs.

bivalves. A group of mollusks, each with a shell in two parts joined by a hinge. In bivalves, the plane of symmetry lies between the two valves (i.e., the valves are mirror images).

blastoids. An extinct group of echinoderms, each with a bud-like body that was attached to the seafloor by a short stem.

body fossils. The actual remains of organisms, such as shells or bones.

brachiopods. A group of invertebrates, each with a shell in two parts joined by a hinge. In brachiopods, the plane of symmetry divides each valve in two.

bryozoans. A group of invertebrates that live in colonies and have skeletons made up of many small boxes or tubes.

caenogastropods. A diverse group of gastropods that includes many familiar types such as periwinkles, whelks, and cowries.

carbonization. The fossilization process by which a creature is preserved as a thin film of carbon.

cast. A fossil formed when the void left by a dissolved shell is filled in with other minerals.

cephalon. The headshield of a trilobite; made of fused segments.

cephalopods. A group of mollusks, each with a large head and mobile tentacles. Most have a single shell divided into chambers. Nautiloids, ammonoids, and belemnoids are types of cephalopods.

ceratites. A group of ammonoids with sutures showing rounded saddles and frilly lobes.

chelicerae. Pincers located in front of the mouth in some arthropods.

chelicerates. A group of arthropods possessing chelicerae. Includes spiders, scorpions, horseshoe crabs, and the extinct eurypterids.

chondrichthyans. A group of fishes with skeletons made of cartilage, not bone. Includes sharks and rays.

chronology. The determination of dates and the sequence of events; the arrangement of events in time.

class. A taxonomic group containing one or more orders.

coccolithophores. A group of single-celled microorganisms, each with a skeleton made of calcium carbonate plates. Their shells are important components of chalky sediments.

comets. Small solar-system objects composed mainly of ice, dust, and rocky material; they orbit the sun.

conglomerate. A coarse-grained sedimentary rock composed of rounded fragments of pre-existing rocks and held together by a fine-grained matrix of sediment or mineral cement.

conifers. A group of gymnosperms with seed-bearing cones and needle-like or scale-like leaves.

contact metamorphism. The solid-state alteration of rocks by the heat of a nearby igneous intrusion.

corals. A group of invertebrates in which the individuals build calcium carbonate skeletons around themselves. Corals may be solitary or live in a colony.

Cordaitales. An extinct group of conifer-like gymnosperms with strap-like leaves.

core. The central part of the earth; thought to consist mostly of iron and nickel.

Corynexochida. A diverse group of trilobites, often spiny, each with a large pygidium and seven or eight segments in the thorax.

cratons. Large, stable pieces of continental crust which are usually made of igneous and/or metamorphic rocks, sometimes with a thin veneer of sedimentary rocks.

Creation. The creation of the universe, the earth, and its living things by God in the space of six days.

creationist. Relating to the model of origins that assumes that God created the universe, the earth, and its living things just as the Bible describes.

crinoids. A group of echinoderms, each with a cup-shaped body and branching arms; typically attached to the seafloor by a long stem. Commonly known as sea lilies.

crust. The thin outer layer of the earth averaging about 6 miles (10 kilometers) thick in the oceans and up to 30 miles (50 kilometers) thick on the continents.

crustaceans. A diverse group of arthropods including crabs, lobsters, shrimps, ostracods, and barnacles.

cyanobacteria. A group of single-celled microorganisms that are capable of photosynthesis.

cycads. A group of gymnosperms, each with a woody trunk, fern-like leaves, and large cones.

dendroids. A group of graptolites with complex, bushy skeletons that were mostly attached to the seafloor.

denticles. In sharks, small, tooth-like scales.

diapsids. A group of tetrapods with two openings in the skull behind the eye. Snakes, crocodiles, and dinosaurs are types of diapsids.

diatoms. A group of single-celled microorganisms with silica shells consisting of two overlapping valves.

disjunct ranges. Distributions in which members of a plant or animal group are found on opposite sides of a geographical barrier (such as an ocean), but nowhere else.

diversification. A rapid process by which a baramin gives rise to new species and varieties.

echinoderms. A group of invertebrates with spiny skin, tube feet, and radial symmetry. Sea urchins, starfish, and sea lilies are types of echinoderms.

echinoids. A group of echinoderms with round or heart-shaped skeletons. Commonly known as sea urchins.

ecological zonation theory. A creationist theory that explains the succession of fossil organisms as being a result of the ecological distribution of organisms before the global Flood, rather than evolution over long ages.

Ediacarans. A diverse and widespread assemblage of soft-bodied organisms with quilted bodies ranging in form from flat discs to domes to fronds.

edrioasteroids. An extinct group of echinoderms resembling a starfish wrapped around a ball or disc.

enantiornithines. An extinct group of birds, most of which were toothed.

entombment. The fossilization process by which creatures are preserved when they are trapped in tree sap, tar, or ice.

erosion. The wearing away of the earth's crust by the action of water, wind, or glacial ice.

eurypterids. An extinct group of chelicerates resembling water scorpions.

evolutionary. Relating to the model of origins that assumes that all organisms on earth descended from a common ancestor by the natural processes of variation and selection.

Fall. The original sin of the first man, Adam, which brought death upon the creation, including all of Adam's descendants.

family. A taxonomic group containing one or more genera.

ferns. A group of plants with frond-like leaves and which reproduce by spores released from the underside of the fronds.

fishes. A group of vertebrates with gills and fins that live wholly or mostly in water.

floating forest. A unique biome of aquatic and semi-aquatic plants thought to have extended over the deep pre-Flood ocean. In creationist theory, it was broken up and buried during the Flood, producing many of the world's coal layers.

Flood. The worldwide watery judgement in the days of Noah that destroyed the pre-Flood world.

foraminifera. A group of single-celled microorganisms with chambered shells. Their shells are important components of many deep-sea sediments.

fossilization. The set of processes by which an organism becomes a fossil.

fossil record. The total of all fossils that have been discovered, and the information that can be derived from them.

fossils. The remains or traces of animals, plants, and other organisms that have been preserved in the earth's crust.

fungi. A group of spore-producing organisms that feed on organic matter.

fusulinids. An extinct group of foraminifera, whose shells are an important component of some limestones.

gastropods. A group of mollusks, each with a well-defined head and a large foot. Most have a single coiled shell. Snails and slugs are types of gastropods.

genus (*pl.* genera). A taxonomic group containing one or more species.

geological column. An idealized representation of the global sequence of rock formations, with the first layers to be deposited at the bottom and the last to be deposited at the top. In conventional theory, the column built up over hundreds of millions of years. In creationist theory, the column built up over thousands of years, much of it during the single year of the Flood.

geology. The scientific study of the origin, history, and structure of the earth.

ginkgos. A group of gymnosperms with fan-shaped leaves. The group now has only one living member (the maidenhair tree) but was more diverse in the past.

glacial. A cold period during which ice sheets build up. Also referred to as an ice age.

gneiss. A coarse-grained metamorphic rock characterized by alternating bands of light and dark minerals.

goniatites. A group of ammonoids with jagged sutures.

granite. A coarse-grained igneous rock consisting of interlocking crystals of the minerals quartz, feldspar, and mica.

graptolites. An extinct group of invertebrates that lived in branching colonies. The individual animals were housed in cups arranged along the branches.

graptoloids. A group of graptolites with simple skeletons, mostly free-floating in habit.

gymnosperms. A diverse group of plants with unprotected seeds. Seed ferns, cycads, ginkgos, and conifers are types of gymnosperms.

Harpetida. A group of trilobites, each with a wide, flat fringe to the cephalon, small eyes, a short pygidium, and twelve or more segments in the thorax.

heterobranchs. A group of gastropods that includes bubble shells, high-spired nerinoids, sea slugs, and many freshwater and land snails.

heterodonts. A group of mostly burrowing bivalves, including many familiar types such as cockles, clams, and razor shells.

heteromorphs. In ammonoids, species that had irregularly coiled shells.

holochroal. Compound eyes in which many lenses are covered by a single corneal membrane.

ice age. The period of time that saw the build-up of ice sheets on the mid- to high-latitude continents. In conventional theory, there were numerous, successive ice ages during the last two million years. In creationist theory, there was one rapid ice advance shortly after the global Flood.

igneous rocks. Rocks that were formed when hot magma cooled below ground, or at the earth's surface when it was erupted as lavas and ashes.

inarticulates. A group of brachiopods that lack a toothed hinge.

insects. A diverse group of arthropods, each with six legs and usually one or two pairs of wings.

invertebrates. Animals without a vertebral column.

kingdom. The highest and broadest category in the taxonomical hierarchy (kingdom, phylum, class, order, family, genus, species).

Lichida. A group of trilobites, each with a broad central part of the cephalon, a large pygidium, and eight to thirteen segments in the thorax.

lissamphibia. A group of amphibians that includes modern types such as frogs, toads, newts, salamanders, and the limbless caecilians.

lycopsids. A group of plants including clubmosses and scale trees.

mammals. A group of tetrapods characterized by hair, milk secretion, and (usually) the birth of live young.

mantle. The part of the earth that lies directly below the crust and above the core. The earth's mantle extends from about 6-30 miles (10-50 kilometers) to 1,800 miles (2,900 kilometers) depth.

marsupials. A group of mammals in which the young are born in an undeveloped state and are sheltered in a pouch. Includes kangaroos, wombats, and possums of Australia and opossums of the Americas.

metamorphic rocks. Rocks that were formed by the solid-state recrystallization of existing rocks without melting. Elevated temperatures and/or pressures cause new minerals to grow, without a change in chemical composition.

microfossils. Fossilized organisms, or parts of organisms, of microscopic size, including foraminifera, radiolaria, diatoms, coccolithophores, pollen, and spores.

mobile genetic elements. Mobile pieces of DNA that may originally have been designed to co-operate with the genes of living organisms to produce rapid biological change.

mold. A fossil formed when a shell dissolves, leaving an impression of its inside (internal mold) or outside (external mold).

mollusks. A group of invertebrates with unsegmented bodies and usually enclosed in a shell. Bivalves, gastropods, and cephalopods are types of mollusks.

monotremes. A group of mammals that lay eggs. Includes platypuses and echidnas of Australia and New Guinea.

morphology. The form, shape, or structure of an organism.

myriapods. A group of arthropods with elongated bodies and numerous leg-bearing segments. Includes centipedes and millipedes.

nautiloids. A group of cephalopod mollusks, some with straight shells and others with coiled shells. The group is now represented by only a handful of species but was more diverse in the past.

neognaths. A very diverse group of birds that includes most of the living species.

neritimorphs. A group of gastropods with globular shells.

nested hierarchy. A hierarchical classification scheme involving levels which consist of, and contain, lower levels.

nummulitids. A group of unusually large foraminifera with numerous coils subdivided into chambers.

ophiuroids. A group of echinoderms, each with a central disc and long, thin arms. Commonly known as brittlestars.

order. A taxonomic group containing one or more families.

orthids. An extinct group of brachiopods, each with a short hinge line, one valve flatter than the other, and a semicircular shape in outline.

osteichthyans. A group of fishes with skeletons made of bone. Includes sarcopterygians (lobe fins) and actinopterygians (ray fins).

ostracoderms. An extinct group of jawless fishes, many with bony head shields.

ostracods. A group of crustaceans in which the body is enclosed between two bean-shaped valves.

palaeognaths. A group of mostly flightless birds including living forms such as cassowaries, emus, ostriches, kiwis, and rheas.

paleontology. The scientific study of fossilized organisms.

palynology. The scientific study of spores and pollen.

Pangea. The reconstructed supercontinent that may have formed and broken up during the Flood.

patellogastropods. A group of gastropods with conical shells and wide openings. Includes limpets.

pedicle. In brachiopods, the fleshy stalk by which the animal attaches to the seafloor.

pentamerids. An extinct group of brachiopods, each with a short hinge line and a typically five-sided shape in outline.

perissodactyls. A group of hoofed mammals, each with an odd number of toes. Horses, tapirs, and rhinoceroses are types of perissodactyls.

petrification. The fossilization process that involves the partial replacement or impregnation of materials such as bone or wood by other minerals.

Phacopida. A large and varied group of trilobites, each with eight to nineteen segments in the thorax and a small pygidium. Includes those with schizochroal eyes.

phragmocone. The chambered portion of a cephalopod shell which may be straight, curved, or coiled.

phylum (*pl.* phyla). A taxonomic group consisting of one or more classes of organisms.

placentals. A group of mammals in which the embryo develops in the mother's uterus, attached by a placenta. Most living mammals are placentals.

placoderms. An extinct group of jawed fishes which were typically heavily armored.

planktonic. Floating or drifting passively in the surface waters of seas or lakes.

plants. A large and diverse group of organisms that typically grow in a permanent site, draw water and inorganic matter through roots, and harness sunlight to produce nutrients.

pleura (*sing.* pleuron). The lateral lobes of a trilobite.

productids. An extinct group of large brachiopods, each with a long hinge line and thick, often spiny, shells.

Proetida. A group of small trilobites, each with a small pygidium and eight to ten segments in the thorax.

proostracum. The forward-projecting hood of the phragmocone in belemnoids.

protobranchs. A group of bivalves with small shells usually lacking ornamentation. Commonly known as nut shells.

psilopsids. Vascular plants with no true leaves, roots, or seeds.

pteriomorphs. A group of bivalves that live on the surface, often in dense colonies. Includes familiar types such as mussels, oysters, and scallops.

Ptychopariida. A large and diverse group of trilobites, each with a typically small pygidium and eight or more segments in the thorax.

pygidium. The tailshield of a trilobite; made of fused segments.

radiolaria. A group of single-celled microorganisms with glassy shells made of silica. Their shells are important components of many deep-sea sediments.

radiometric dating. A set of methods for estimating the ages of rocks and minerals using the relative abundances of radioactive and stable isotopes of certain elements.

radula. The tooth-like parts of a mollusk used to scrape food particles from a surface and draw them into the mouth.

reconstruction. An impression or model of an event or entity from the past formed from the available evidence.

Redlichiida. A group of spiny trilobites, each with large eyes, a tiny pygidium, and many segments in the thorax.

reef. A mound-like structure built by marine organisms, especially corals, and consisting largely of their remains.

regional metamorphism. The solid-state alteration of rocks over broad areas of the earth's crust, usually involving elevated temperatures and pressures.

replacement. The fossilization process by which materials such as shell or bone are dissolved and replaced atom by atom with other minerals.

reptiles. A group of tetrapods with dry, scaly skin that usually lay soft-shelled eggs on land.

rhabdosome. A graptolite colony.

rhynchonellids. A group of brachiopods, each with a short hinge line and strongly ribbed shells.

Rodinia. The reconstructed supercontinent that may represent the arrangement of the continents before the Flood.

rostrum. The internal shell of a belemnoid, sometimes called the guard.

rugose corals. An extinct group of corals with bilaterally symmetrical skeletons and distinct radial walls. Some lived singly and others formed colonies.

runaway subduction. The process by which the ocean crust sank into the earth's interior at rates of meters-per-second, initiating the catastrophic break-up of the continents during the global Flood.

sarcopterygians. A group of bony fishes in which the fins are borne on a fleshy lobe attached to the body. Commonly known as lobe fins.

schizochroal. Compound eyes in which each lens has its own corneal membrane.

scleractinian corals. A group of corals with radially symmetrical skeletons and distinct radial walls. Some live singly and others form colonies.

sedimentary rocks. Rocks that were formed by the erosion of earlier rocks, the precipitation of dissolved chemicals, or the accumulation of organisms. The resulting sediment is typically deposited in layers.

seed ferns. An extinct group of gymnosperms with fern-like fronds bearing seeds.

species. A population of interbreeding organisms that is reproductively isolated from other such populations.

sphenopsids. A group of plants, including horsetails, characterized by jointed stems with whorls of leaves at the joints.

spiriferids. An extinct group of brachiopods, each with a long hinge line and a ribbed shell with tapering "wings".

sponges. A group of invertebrates with skeletons made of needle-like crystals of calcium carbonate or silica.

spreading ridges. Plate boundaries at which two tectonic plates are separating.

Stigmaria. The roots of lycopsids.

stipe. The individual stalks or branches that make up a graptolite rhabdosome.

strata (*sing.* stratum). Sedimentary rock layers.

stromatolites. Layered, dome-shaped structures built by colonies of sticky cyanobacteria that trap sediments.

subduction zones. Plate boundaries at which one tectonic plate descends beneath another and is consumed in the earth's mantle.

supercontinent. A large landmass that broke apart to form several continents.

sutures. In nautiloids and ammonoids, the distinctive lines that mark where the partitions separating each chamber meet the external shell. The parts that point forwards are called saddles; the parts that point backwards are called lobes.

synapsids. A group of tetrapods with one opening in the skull behind the eye. Pelycosaurs and therapsids are types of synapsids.

tabulate corals. An extinct group of corals with radially symmetrical skeletons and indistinct or absent radial walls. All members of this group formed colonies.

taxonomy. The science of naming and classifying organisms.

tectonic plates. Rigid slabs of the earth's crust and upper mantle, typically hundreds to thousands of miles across. A continental plate carries a continent, but also has oceanic crust. An oceanic plate has no continental material.

terebratulids. A group of brachiopods, each with a short hinge line and smooth, bulbous shells.

tetrapods. A group of vertebrates possessing limbs with digits (not fins). Includes the amphibians, reptiles, birds, and mammals.

theca (*pl.* thecae). The small cup in which an individual graptolite animal is housed.

thorax. In arthropods, the part of the body behind the head, each segment of which bears legs or wings.

trace fossils. Fossils that result from the activity of living organisms, such as burrows, tracks, and trails.

transform faults. Plate boundaries at which the tectonic plates are slipping past one another, and the fault movement is mostly horizontal.

tree ferns. A group of plants, each with frond-like leaves sprouting from a trunk.

trilobites. An extinct group of arthropods, each with a body divided lengthways into three lobes.

ungulates. A diverse group of mammals with hooves. Includes perissodactyls (odd-toed ungulates) and artiodactyls (even-toed ungulates).

vertebrates. Animals with a vertebral column.

vetigastropods. A group of gastropods, each with a slit-like opening in the outer lip of the shell. Commonly known as slit shells.

zooids. The individual animals that make up a colonial organism such as a graptolite.

INDEX Italic page numbers indicate associated illustrations.

U

W